Springer Earth System Sciences

More information about this series at http://www.springer.com/series/10178

Beatriz L. L. Coira · Clara Eugenia Cisterna

Textures, Structures and Processes of Volcanic Successions

Examples from Southern Central Andes (Northwestern Argentina, 22°–28°S)

Springer

Beatriz L. L. Coira
Instituto de Ecorregiones Andinas
(INECOA), CONICET-Universidad
Nacional de Jujuy
San Salvador de Jujuy, Argentina

Clara Eugenia Cisterna
Facultad de Ciencias Naturales e Instituto
Miguel Lillo
Universidad Nacional de Tucumán –
CONICET
Tucumán, Argentina

ISSN 2197-9596 ISSN 2197-960X (electronic)
Springer Earth System Sciences
ISBN 978-3-030-52009-0 ISBN 978-3-030-52010-6 (eBook)
https://doi.org/10.1007/978-3-030-52010-6

This Springer imprint is published by the registered company Springer Nature Switzerland AG
The registered company address is: Gewerbestrasse 11, 6330 Cham, Switzerland

To Alejandro José Perez inseparable companion of Andes adventures alongside

To Magdalena Koukharsky with whom we imagine this book

Preface

The Central Andes provides an excellent natural laboratory for studying the processes associated with the subduction of oceanic crust beneath a continental margin.

The onset of Andean magmatic cycle is considered to have taken place in Late Triassic–Early Jurassic Period (Dalziel 1986) although subduction had already commenced in the Late Carboniferous Period after the passive margin configuration. The volcanic activity has been essentially continuous from that time to the present days (Coira et al. 1982; Jordan et al. 1983, 1997; Ramos 1988; Mpodozis and Ramos 1989; Kay et al. 1991, 1999; Kay and Abbruzzi 1996) offering examples whose analysis open new perspectives in the understanding of volcanic processes linked to active continental margins as the Central Andes. In the same regional context are not less important the ancient volcanic records forming the Ordovician sequences that outcrop along the Puna and Famatina System. Their study gives the opportunity to understand the evolution of the southwestern active margin of Gondwana during the Paleozoic.

As a result of our research carried out over the last 40 years on volcanism (Cenozoic to Paleozoic) of the Southern Central Andes, especially of the northwestern Argentina, we have collected illustrative material, on both modern and ancient volcanic successions, which show a great diversity of types of rocks and processes, reflecting the particular conditions resulting from the conjunction of lithospheric structures, tectonic history, and magmatism, during the evolution of the Andean continental margin. We believe that taking into account together in a book the *Textures, structures and processes of volcanic successions: examples from Southern Central Andes (Northwestern Argentina)*—will allow to get through the Central Andean volcanism from the Lower Paleozoic to the Cenozoic times in the framework of their processes, eruptive mechanisms, and geodynamic conditions, awakening the interest of volcanologists and specialists in Earth Sciences.

For the purposes of the treatment of the different types of volcanic successions to be considered in the book, the more representative volcanism records of the Southern Central Andes have been selected that took place during the Cenozoic as caldera complexes, composite volcanoes (stratovolcanoes), silicic dome complexes, and in monogenetic mafic centers as well as in the Paleozoic times as submarine–

subaerial volcanism. The analysis of the Paleozoic ancient volcanic successions, in which recognition of the original rock type can be difficult by the effects of deformation, metamorphism, and alteration, is a subject of particular attention.

In the present book, the description of distinctive cases of Southern Central Andes volcanism provides to the readers useful tools to be applied in the analysis and interpretation of volcanic successions linked to the evolution of active continental margin.

San Salvador de Jujuy, Argentina — Beatriz L. L. Coira
Tucumán, Argentina — Clara Eugenia Cisterna

References

Coira BL, Davidson JD, Mpodozis C, Ramos VA (1982) Magmatic evolution of the Andes of northern Argentina and Chile. Earth Sci Rev 18:303–332

Dalziel IWD (1986) Collision and cordilleran orogenesis: an Andean perspective, Geological Society, London, Special Publications 19:389–404

Jordan TE, Isacks BL, Ramos VA, Allmendinger RW (1983) Mountain building in the Central Andes. Episodes 20–26

Jordan TE, Reynolds JH, Erikson JP (1997) Variability in age of initial shortening and up-lift in the central Andes, 16-33°30 S. In: Ruddiman W (ed) Tectonic Uplift and Climate Change, Plenum, New York, pp 41–61

Kay SM, Abbruzzi JM (1996) Magmatic evidence for Neogene lithospheric evolution of the central Andean "flat-slab" between 30° and 32°S. Tectonophysics 259:15–28

Kay SM, Mpodozis C, Coira B (1999) Neogene Magmatism, tectonism and mineral deposits of the Central Andes (22°–33°S latitude). In Skinner BJ (eds) Geology and ore deposits of the Central Andes. Society of Economic Geology Special Publication, vol 7, pp 27–59

Kay SM, Mpodozis C, Ramos VA, Munizaga F (1991) Magma source regions for mid to late Tertiary volcanic rocks and erupted over a shallowing subduction zone and through a thickening crust in the main Andean Cordillera (28-33"s). In: Harmon RS, Rapela CW (eds) Andean Magmatism and its Tectonic Setting. Geol Soc Am Spec Pap 265:113–137

Mpodozis C, Ramos VA (1989) The Andes of Chile and Argentina. In: Ericksen GE, Cañas Pinochet MT, Reinemud JA (eds). Geology of the Andes and its relation to hydrocarbon and mineral resources. Circum Pacific Council for Energy and Mineral resources Earth Science Series 11:59–90

Ramos VA (1988) Tectonics of the Late Proterozoic—Early Paleozoic: a collisional history of Southern South America. Episodes 11(3):168–174

Acknowledgments

This book results from many years of working in the Southern Central Andes and the contributions of many colleagues. Beatriz L. L. Coira particularly thank Suzanne Kay, Magdalena Koukharsky, José Viramonte, Pablo Caffe, Michael Ort, Miguel Soler, Ivan Petrinovic, Ruben Somoza, Diego Frachia, Gustavo Rodríguez, and Belén Pérez for fruitful contributions and discussions on the uplift, magmatic, and tectonic processes in the Southern Central Andes and also like to make a special mention to the always present collaboration and accompaniment of Alejandro José Pérez during the different stages of research carried out over the years.

Likewise, the assistance of Edgardo González in preparing the book should, in particular, be appreciated, especially for the careful and excellent processing of the illustrations, photographs, and maps, and revising drafts.

The institutions Consejo Nacional de Investigaciones Científicas y Técnicas (CONICET), Universidad Nacional de Jujuy, and Universidad Nacional de Tucumán contributed to make this book possible giving support to the researches carried out by the authors.

About This Book

Textures, structures, and processes go together in the understanding of volcanism and its geodynamic environment. In this book, the attention to achieve this proposal is focused on the analysis of voluminous Cenozoic volcanism which alongside the extreme crustal shortening and thickening have been crucial in the evolution of the Central Andes. In the same sense, the volcanic successions of the wide and extended Lower Paleozoic magmatic belt (Famatinian arc is exposed for ~2000 km along the Southern Central Andes) also provide excellent examples for studying ancient volcanic–volcaniclastic and volcanogenic records, critical in reconstructing eruptive and depositional processes in the evolution of the Famatinia Belt, with profoundly influence in the lithospheric architecture of the modern Central Andes.

For the purposes of the treatment of the different types of volcanic successions to be considered in the book, the more representative, well-exposed, and preserved volcanic records of the Southern Central Andes have been selected. In the case of the Cenozoic volcanism included in Part I are analyzed characteristic volcanic edifices of the region that show a complete register of the activity and represent volcano-tectonic events decisive in the geodynamic evolution of these belts. Among them are selected caldera complexes, composite volcanoes (stratovolcanoes), silicic dome complexes, and monogenetic mafic centers, which are treated in independent chapters. The main topics considered in these chapters are types of volcanic buildings and their structure and morphometry, characterization of volcanic deposits in their different types, pyroclastic, lavic or syn-volcanic intrusive, and their mineralogical and geochemical characteristics. Are also taking into account facies analysis as well as differentiation of primary volcanic structures and textures regarding the secondary ones, resulting from the fragmentation, remobilization, particularly reworking or sorting, related to the type of transport from the source or affected by superimposition of tectonic deformation and/or hydrothermal alteration.

The Paleozoic volcanic–volcaniclastic and volcanogenic records (Ordovician and Carboniferous) in submarine–subaerial environments are analyzed in Part II considering well-represented and exposed successions from Famatina System and Puna. Observation and detailed analysis of their facies, provide the tools to recognize original textures and structures as well as syn-volcanic and post-volcanic modifications, giving clues for the difficult task to reconstruct eruptive and depositional processes in the case of ancient successions.

Each of volcanic successions treated include illustrative examples accompanied by maps, stratigraphic sections, petrographic and geochemical data, and photographs and microphotographs of structures and textures. These examples will allow the reader to know the different typologies that characterize the volcanism, environment of emplacement, and tectonic setting linked to active continental margins as the Southern Central Andes.

Contents

Part I
Cenozoic Volcanism

In the Southern Central Andes, the Cenozoic volcanism reach great profusion mainly during the Miocene–Pliocene, covering a large area of the region determining a strong control in the geological features of the region. The most prominent feature in the Central Andes is the extensive Puna–Altiplano plateau which after the Tibetan plateau is the world's highest (average elevation 3700 meters) and largest (700 by 200 km) continental plateau.

Voluminous mafic dacitic to rhyodacitic ignimbrite fields linked to giant calderas are distinctive features of the late Miocene to Quaternary magmatic records of the central Andean Altiplano–Puna plateau. The magmas erupted through a compressional thickened crust over a generally shallow, but variably dipping subducting Nazca plate.

In the analysis of the Central Andean Altiplano–Puna plateau Cenozoic volcanism, the attention in this contribution will be on the Puna plateau (22°S–28°S). The Northern Puna (~ 22°S–24°S) voluminous ignimbrites (> 8000 km^3) and a narrower Subandean fold-thrust belt, that gives way southward to a thick-skinned thrust belt, are distinctive. Their evolution can be modeled by an early Miocene amagmatic flat slab whose steepening after 16 Ma leads to mantle melting that culminated in widespread ignimbrite eruptions beginning at 10 Ma, peaking in the back-arc at ~ 8.5–6 Ma, restricted to the near arc by 4.5 Ma and ending by 3 Ma (de Silva et al. 2006; Kay and Coira, 2009; Kay et al. 2010).

The giant calderas associated with voluminous ignimbrites alongside the compound volcanoes or stratovolcanoes constitute the most conspicuous representatives of the Northern Puna volcanism. Meanwhile silicic dome complexes and monogenetic to polygenetic mafic centers have a more restricted representation.

In the Southern Puna (~ 24°S to ~28°S) is notable an eastward frontal arc migration at 8–3 Ma, an intra-plateau basin bounded by high ranges, long-lived Miocene stratovolcanic–dome complexes, voluminous 6–2 Ma ignimbrites, 7–0 Ma back-arc mafic flows, and the latest Miocene uplift of the reverse faulted Sierras Pampeanas ranges to the east. The evolution can be modeled by a moderately shallow slab producing widespread volcanism with subsequent steepening by 6 Ma leading to delamination of dense lithosphere culminating in the eruption of the voluminous

Cerro Galán Ignimbrite at 2 Ma (Kay and Coira, 2009; Kay et al. 2010; Risse et al. 2013).

In this Part I will be analyzed, through selected examples, morpho-structural and compositional characteristics, distributions and dimensions of Central Andean Puna plateau Cenozoic volcanic centers, studying magmatic processes and the geodynamic conditions during their evolution.

References

de Silva SL, Zandt G, Trumbull RB, Viramonte JG, Salas G, Jimenez M (2006) Large-scale silicic volcanism in the Central Andes — a tectonomagmatic perspective. In: Troise C, de Natale G, Kilburn CRJ (eds) Mechanisms of activity and unrest at large calderas. Geological Society of London SpecialPublication, 269, pp 47–63

Kay SM, Coira B (2009) Shallowing and steepening subduction zones, continental lithospheric loss, magmatism and crustal flow under the Central Andean Altiplano-Puna plateau. In Kay SM, Ramos VA, Dickinson WR (eds), Backbone of the Americas: Shallow Subduction, Plateau Uplift and Ridge Collision, The Geological Society of America Memoir 204:229–258

Kay S, Coira B, Caffe P, Chen C (2010) Regional chemical diversity, crustal and mantle sources and evolution of Central Andean Puna plateau ignimbrites. J Volcanol Geotherm Res 198: 81–110

Risse A, Trumbull RB, Kay SM, Coira B, Romer RL (2013) Multi-stage evolution of late Neogene mantle-derived magmas from the central Andes back-arc in the Southern Puna Plateau of Argentina. J Petrol 54(10):1963–1995

Chapter 1
Calderas

Abstract The large Miocene to Quaternary silicic ignimbrite fields in the Central Andes form one of the largest ignimbrite provinces on Earth, related to giant calderas that occur in the back-arc of the modern Central Volcanic Zone and extend from 14°S to 28°S. The Puna–Altiplano (22°S to 28ª S) dacitic–rhyolitic ignimbrites, greater than 600–1000 km^3, are related to giant calderas, with diameters up to 60 × 35 km. Their collapse styles are "piecemeal", "trapdoor", and "down-sag" and their postulated origin ranges from failure of magma chamber roofs to the overpressure type. These ignimbrites are massive, lack fall deposits, and are crystal rich, with quartz, plagioclase, biotite, Fe Ti oxides, and apatite and titanite accessories. Amphibole occurs in ignimbrites near the arc, sanidine in those with >69% SiO_2, and pyroxenes when SiO_2 <67%. They belong to high-K calc-alkaline ignimbrites with 64–72% SiO_2 and LILE and LREE enrichment relative to the HREE, negative spikes in Nb, Ta, and Ti, typical of arcs and positive spikes in Th and U of continental crust. Its generation is associated with episodes of slab steepening during which mantle magma generation is triggered by mantle wedge melting under a hydrated lithosphere. Episodes of delamination of the crust and mantle lithosphere are also associated.

Keywords Cenozoic · Silicic ignimbrites · Puna · Giant calderas

1.1 Introduction

The large Miocene to Quaternary silicic ignimbrite fields in the Central Andes form one of the largest ignimbrite provinces on Earth. Those voluminous dacitic to rhyodacitic ignimbrites, related to giant calderas, occur in the back-arc of the modern Central Volcanic Zone and extend from 14°S to 28°S. Among them are, on the Altiplano, north of 21°S, the largest ignimbrite centers like the Macusani in Peru (10–6.7 Ma, see Cheilletz et al. 1992) and the Morococala (~8–6 Ma, see Morgan et al. 1998) and Kari-Kari–Los Frailes (see Fig. 1.1) ignimbrite complexes (17–2 Ma, see Schneider 1987) in Bolivia. To the south are distributed a profuse number of ignimbrite centers (5.6–3.2 Ma) like La Pacana, Guacha (Lindsay et al. 2001), Pastos Grandes (Pacheco and Ramirez 1997), included in the Altiplano–Puna Volcanic

B. L. L. Coira and C. E. Cisterna, *Textures, Structures and Processes of Volcanic Successions*, Springer Earth System Sciences,
https://doi.org/10.1007/978-3-030-52010-6_1

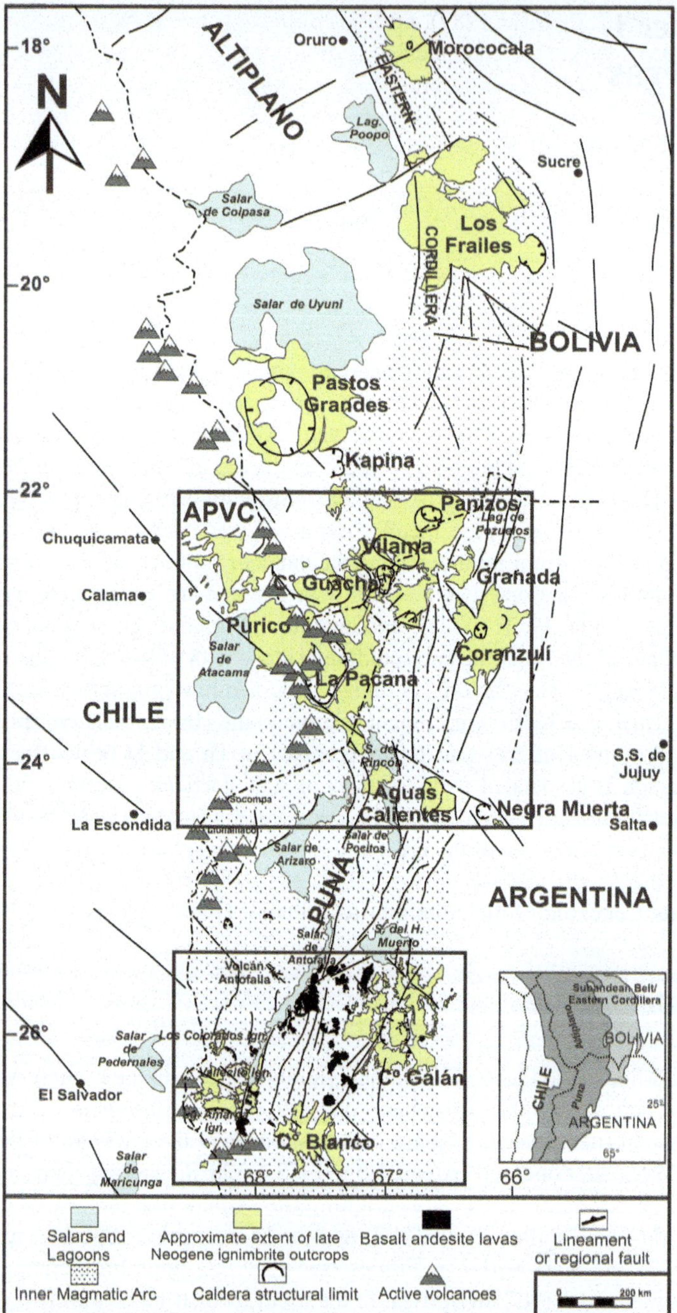

Fig. 1.1 Distribution of main late Neogene Altiplano–Puna calderas and ignimbrite fields (modified from Kay et al. 2010). The examples of ignimbrite and volcanic centers analyzed in this chapter are located in the boxes

Complex (APVC, de Silva 1989), near the Bolivian–Chilean and Argentine border Further east, in the back-arc giant calderas like Vilama (Coira et al. 2004; Soler et al. 2007), Panizos (Ort 1993), and Coranzuli (Seggiaro 1994; Seggiaro et al. 1987, 2019) erupted voluminous ~10–6 Ma ignimbrites (Fig. 1.2). The largest centers south of 24°S include Aguas Calientes (Petrinovic 1999; Petrinovic et al. 2010), Cerro Galán (Folkes et al. 2011a, b; Kay et al. 2011), Laguna Amarga (Seggiaro et al. 1999), and Cerro Blanco (Arnosio et al. 2005; Báez et al. 2015) caldera complexes.

The Puna–Altiplano large ignimbrites spread out in a region characterized by a ~50 to 75 km thick crust (e.g., Beck and Zandt 2002; Yuan et al. 2000; McGlashan et al. 2008) with elevation over 3700 m and an active compressional deformational belt to the east. Its generation is temporally associated with episodes of slab steepening during which mantle magma generation is triggered by decompression melting in an expanding mantle wedge under a hydrated lithosphere (see Kay et al. 1999; Kay and Coira 2009). Episodes of delamination of the crust and mantle lithosphere are associated with ignimbrite formation. Temporal and spatial patterns of those ignimbrites are used to argue that the largest volume of ignimbrites occur in the regions over which the most slab steepening and delamination of continental crust and lithosphere have occurred (e.g., Kay et al. 1994; de Silva et al. 2006; Kay and Coira 2009).

The main characteristics of the Altiplano–Puna ignimbrites were described and analyzed through the studies carried out in the distinctive calderas of the region by Sparks et al. (1985); Seggiaro et al. (1987, 2019); de Silva (1989); Francis et al. (1989); Coira et al. (1993); Ort (1993); Coira et al. (1996); Petrinovic et al. (1999); Lindsay et al. (2001); Schmitt et al. (2001); Arnosio et al. (2005); Soler et al. (2007); Guzman 2009; Folkes et al. (2011a, b); Kay et al. (2010, 2011); among others.

In general, the large ignimbrites are massive, show few segregation structures, lack fall deposits, are crystal rich, have relatively uniform crystal sizes, and show little evidence for fluidization. These features have been interpreted to reflect low explosive eruptions that develop collapsing columns of reduced height (boiling over-type columns) and produce viscous pyroclastic flows.

The post-collapse magmatic resurgence is common in those systems. The high viscosity, crystal-rich, low volatile content of those large dacitic magmas have been argued to require external triggers and chamber deflation to initiate the collapse of the roofs that drive and sustain the eruptions (e.g., de Silva et al. 2006; de Silva and Gosnold 2007; Soler et al. 2007; Martí et al. 2009). Those voluminous (100 s to 1000 s km^3) ignimbrites generally shared similar petrographic characteristics. The phenocryst assemblage comprises quartz, plagioclase, biotite, and Fe Ti oxides with accessory apatite and titanite. Amphibole is common in some Northern Puna ignimbrites near the arc, sanidine is found in more silicic ignimbrites (>69% SiO_2), and clinopyroxene and orthopyroxene can occur in ignimbrites with <67% SiO_2. They belong to high-K calc-alkaline dacitic to rhyodacitic ignimbrites (64–72% SiO_2), those that are framed in the "monotonous intermadiate" type defined by Hildreth (1981) and show a broad range of aluminosity with the Northern Puna Miocene back-arc ignimbrites being the most peraluminous, whereas the Northern Puna near-arc ignimbrites are metaluminous and the Southern Puna ones (Cerro

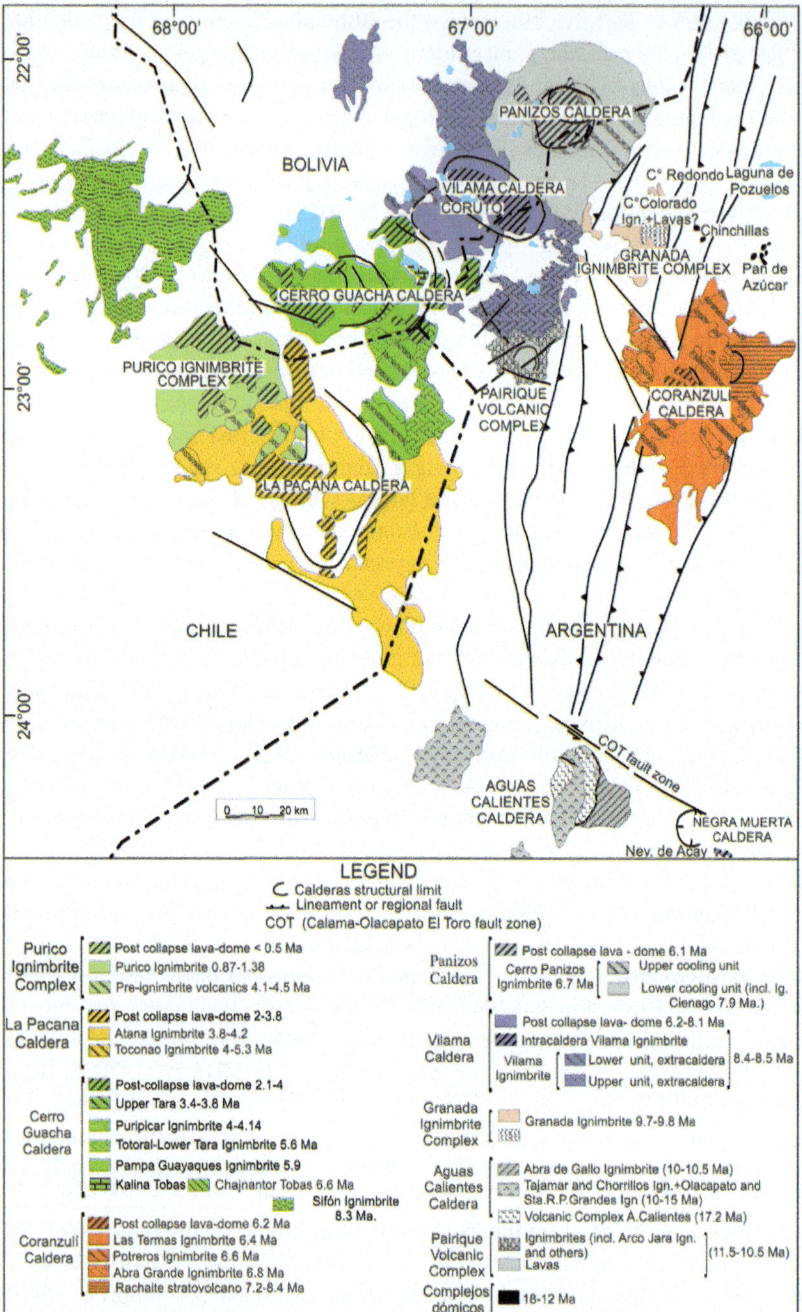

Fig. 1.2 The major calderas and volcanic units of the Northern and Central Puna–Southern Altiplano centers (modified from Kay et al. 2010)

Galán ignimbrites) straddle the metaluminous boundary. Their trace element patterns show LILE and LREE enrichment relative to the HREE; negative spikes in Nb, Ta, and Ti indicating variable degrees of the HFSE depletion typical of arc magmas, and positive spikes in Th and U, typical of continental crustal components (Kay et al. 2010).

In general Northern Puna back-arc ignimbrites are Na_2O poor and have higher La/Ta, La/Yb, and $^{87}Sr/^{86}Sr$ ratios and $\delta_{18}O$ values than near-arc ignimbrites. Southern Puna ignimbrites are generally intermediate in character. These features are compatible with crustal melt contributions being more shale-like in the Northern Puna back-arc, more amphibolite-like near the arc, and more biotite gneiss-like in the Southern Puna. Melting in the deep crust by injection of mantle-derived melts followed by magma rise, accumulation, and evolution at depths of nearly 25–20 km is supported by heavy REE evidence for deep crustal garnet-bearing residues, negative Eu anomalies superimposed on steep REE patterns, calculated bulk Sr distribution coefficients, and Puna seismic images (Kay et al. 2008, 2010, 2011).

The eruption of those voluminous dacitic–rhyolitic ignimbrites (ignimbrite flare-up), greater than 600–1000 km^3, was related to giant calderas with circular to elliptical geometries with variable diameters, the largest being La Pacana (60 × 35 km) and Vilama (40–35 × 18–15 km) and the smallest one Cerro Blanco (6–5 × 5–4 km). The collapse styles most likely are "piecemeal", "trapdoor and "down-sag" (Sparks et al. 1985; Gardeweg and Ramirez 1987; Ort 1993, Folkes et al. 2011a; Lindsay et al. 2001; Soler et al. 2007; Di Fillipo et al. 2008; Guzman and Petrinovic 2010). Most calderas did not have a Plinian eruption prior to its formation, except in cases as some units of Cerro Blanco, Grupo Toconquis, Panizos and Vilama ignimbrites.

De Silva et al. (2006) suggest that the most like trigger for Andean calderas is not produced by magmatic overpressure, but by mechanical failure of magma chamber roofs. Meanwhile Martí et al. 2009; Petrinovic et al. 2010; Guzman and Petrinovic 2010 postulate that those calderas would correspond to the overpressure type with early caldera subsidence, from the beginning of the eruption itself, which would generate eruptions of type "boiling over".

A challenge assumed in this chapter is to analyze the evolutionary characteristics, structures and processes that controlled the generation of those giant calderas, as well as the distinctive features of the voluminous dacitic to rhyodacitic ignimbrite fields of one of the largest ignimbrite provinces on Earth linked to them. Examples of calderas and ignimbrite fields of Northern and Southern Puna will be subject to this analysis.

1.2 Northern Puna Calderas

During the upper Miocene the volcanism reached its climax in the area of the Northern Puna. Large ignimbrite volumes, associated with boilers of gigantic dimensions, were emitted in the last 10 Ma as part of the *Neogene Ignimbrite Province of the Central Andes*. Its evolution took place in the thickened continental crust segment, under an active subduction regime of increasing angle in a compression margin (Coira et al. 1993; Kay et al. 1999) and reflects particular thermo-mechanical conditions of the crust in this sector of the Andes.

In the Northern Puna the ignimbrite episodes had an *early stage* at 10 Ma, which marked the beginning of the large-volume ignimbrite volcanism. In its course there were eruptions of reduced volumes with respect to the subsequent stage. Its emission centers remain hidden below younger volcanic buildings. In lapse 8–4 Ma, a large-scale ignimbrite volcanism was established (*ignimbrite climax*), to which the development of gigantic calderas was linked. Among these calderas are highlighted in the period 8–6 Ma, Vilama, Panizos, Coranzulí and in Bolivian territory, Guacha (Fig. 1.3). During the Pliocene the ignimbrite volcanism migrated to the west in

Fig. 1.3 Major calderas of the Northern Puna analyzed in this chapter

response to the increase in the subduction angle, being represented by the La Pacana caldera and its voluminous ignimbrite units (5–3.4 Ma), as well as by new pyroclastic flows that marked the reactivation for those activity times in the Guacha caldera. The ignimbrite volcanism continued with less volume in that region during the Pleistocene, represented by the Purico Ignimbrite Complex (1.3 Ma) and minor eruptions associated with nested centers within the La Pacana caldera.

1.2.1 Panizos Caldera

The Panizos caldera, located in the southern CVZ at 22°15′S, 66°45′W, (Fig. 1.4) studied in detail by Ort (1993) and Ort et al. (1996), is the source of a large volume dacitic ignimbrite (Cerro Panizos Ignimbrite—CPI) of 650 km^3 of total erupted volume (DRE, minimum).

This Ignimbrite was emplaced at 6.7 Ma and forms a 40 km diameter shield-like plateau with a 1–3° dip radially outward from a 10–15 km diameter cluster of dacitic lava domes. Ort (1993) recognized, as part of the Cerro Panizos Ignimbrite (CPI), a simple cooling unit according to Smith (1960) and, Christiansen (1979), represented by the "Lower Cooling Unit" (LCU) overlain by a compound cooling unit "Upper Cooling Unit" (UCU), separated by bedded pyroclastic deposits (surge deposits). The "Lower Cooling Unit" (LCU) (Fig. 1.5a) is a simple cooling unit with a maximum thickness of 160 m. It is characterized by its richness in crystals, absence of significant variation in pumice and lithic concentrations, variable welding (unwelded to densely welded) (Fig. 1.5b), absence of fall deposits and vapor-phase crystallization at the top, typical of simple cooling unit. Lithics reach significant concentration (>5%) in the upper two meters of LCU (Fig. 1.5c); they could document vent wall collapse or the opening of a new vent. Gas pipes recognized in some sections (Fig. 1.5d) are indicative of fluidization processes during the emplacement. The continuous cooling zonation and the lack of evidence of eruption column fluctuations, such as fallout layers, imply steady eruption of hundreds of cubic kilometers of Lower Cooling Unit material from a single vent or a small cluster of vents (Ort 1993).

The Upper Cooling Unit (a compound ignimbrite cooling unit) is composed of several flow units with variations in total thickness (30–165 m), welding, and lithic content (Fig. 1.5e). Thin-bedded pyroclastic surge are recorded at the base (Fig. 1.5f). Thin, moderately well-sorted pumice-fall layers (Fig. 1.5g) occur at various levels throughout that unit. These characteristics were interpreted by Ort (1993), as a result of an unstable eruption column with the opening of several mouths associated to cauldron collapse. Large lateral and vertical variations in welding, and sectorial distribution of flow units, indicate vent migration during the eruption.

Vapor-phase crystallization is continuous from the upper part of the LCU through surge deposits into the UCU indicating the absence of a significant temporary break among them and pointing to consider that the Cerro Panizos Ignimbrite is a composite sheet (Smith 1960) associated to a single eruptive period, rather than the product of two distinct eruptions (Ort 1993). A 15 km diameter topographic depression, marked

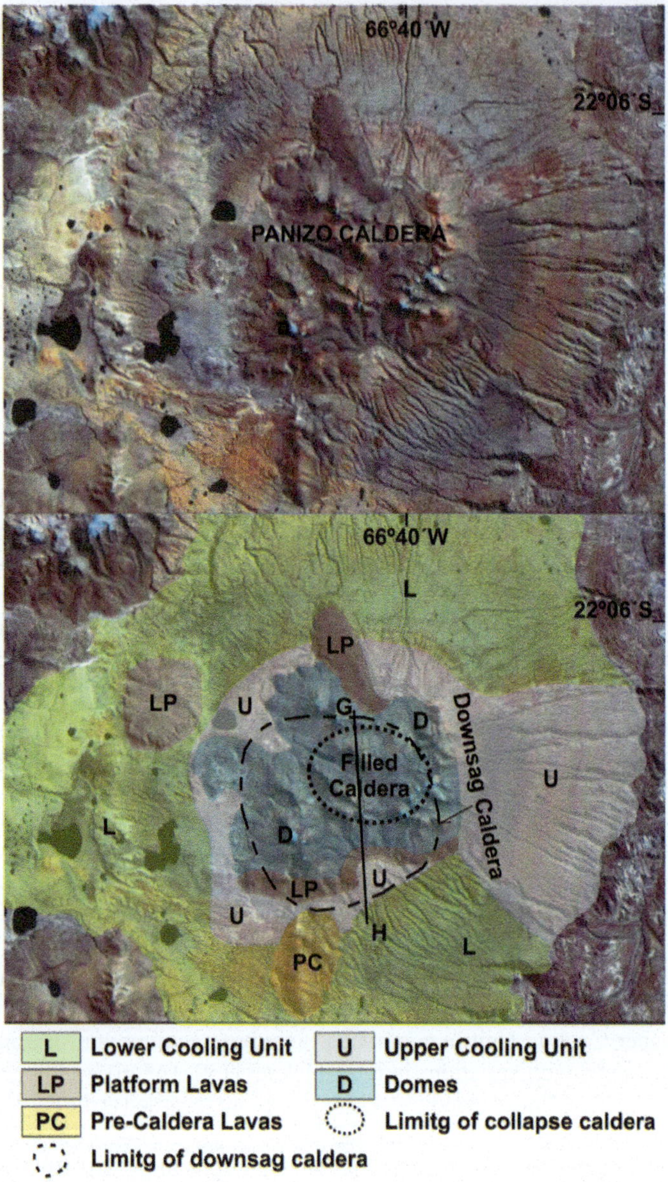

Fig. 1.4 Panizos caldera

Fig. 1.5 Panizos Lower Cooling Unit (LCU): **a** Columnar-jointed section of moderately to incipiently welded ignimbrite. **b** Unwelded ignimbrite rich in pumice and crystals and poor in lithics. **c** Lithics reach significant concentration (20–30%) in the upper two meters of LCU. **d** Gas pipes in LCU, indicative of fluidization processes during the emplacement. Panizos Upper Cooling Unit: **e** composite by several flow units. **f** Thin-bedded pyroclastic surge at the base of the ash flow. **g** Pumice-fall layer moderately sorted

by inward dips of 4°–8° at the cooling units' contact, is centered in the vent area of the LCU. Ort (1993) interpreted that depression as a down-sag caldera formed during the emplacement of the LCU and indicated that the caldera collapse began late in the eruption of the LCU and continued through the emplacement of the UCU close to the radial ring of central domes, and that its formation was associated with the collapse of the caldera and to eruptions from multiple conduits. Post-caldera effusive volcanism (6.1 Ma) largely covered these structures. The lava flows escaped the caldera radially, partly filling the depression of the previous "down-sag" caldera (Ort 1993). The central domes ascended along fractures defined by the collapsed caldera and filled any topographic depression.

1.2.2 Vilama Caldera

The Vilama caldera (22° 24′S, 66° 57′W) was originally studied by Coira et al. (1996), and later Soler (2005) and Soler et al. (2007) continued their research, reinterpreting some of their characteristics in a detailed analysis. Fracchia (2009) studied the post-collapse volcanism of this caldera.

The Vilama Ignimbrite (8.4–8.5 Ma) with a recognized areal extension 4000 km^2, and estimated volume ~1800 km^3 (1400–1000 km^3 in dense rock equivalents DRE) is related to the Vilama caldera. It is roughly rectangular (35–40 km × 15–18 km) in shape and has central coordinates of 22°24′S, 66°57′W and is considered to have formed in a single-stage collapse event (Soler et al. 2007).

Intra- and extra-caldera facies of the Vilama Ignimbrite are recognized (Fig. 1.6). The extra-caldera facies has a mean thickness of ~40 m and is separated into a restricted valley-ponded Lower Cooling Unit (LCU) and a laterally extensive low aspect ratio (LARI) Upper Cooling Unit (UCU).

The Lower Cooling Unit (LCU) is formed by massive ignimbrites, generally unwelded (Fig. 1.7a), although surge (Fig. 1.7b) and fall deposits are sparsely intercalated in some localities. Those ash flows have high to low (20–3%) amounts of pumice and scarce (<1%) lithics. (Fig. 1.7c). The LCU is less regionally extensive than the UCU, missing in places where the UCU directly mantles older units. The total thickness of the LCU is extremely variable ranging from 7 m to more than 110 m. The unwelded LCU rocks show vitroclastic textures, in which highly vesiculated pumice with a crystal content of ~20–25% are immersed in a vitric shard matrix (Fig. 1.7d).

The Upper Cooling Unit (UCU), the most extensive of the Vilama Ignimbrite (VI) outflow, consists of a single ignimbrite sheet, moderate to intense welded (Fig. 1.7e) with lack of segregation structures and fall deposits and with a thickness of 60–20 m. The basis of this unit is massive, without lithics, highly welded with development of "fiamme" with about 50% crystals (Fig. 1.7f) in all its thickness and a subhorizontal jointing, well developed, which is usually coincident with foliation planes (Fig. 1.7g) and characteristic in some vitrophyric sections. Toward upper sectors, welding decreases and the ash flow is moderately rich in fiamme (up

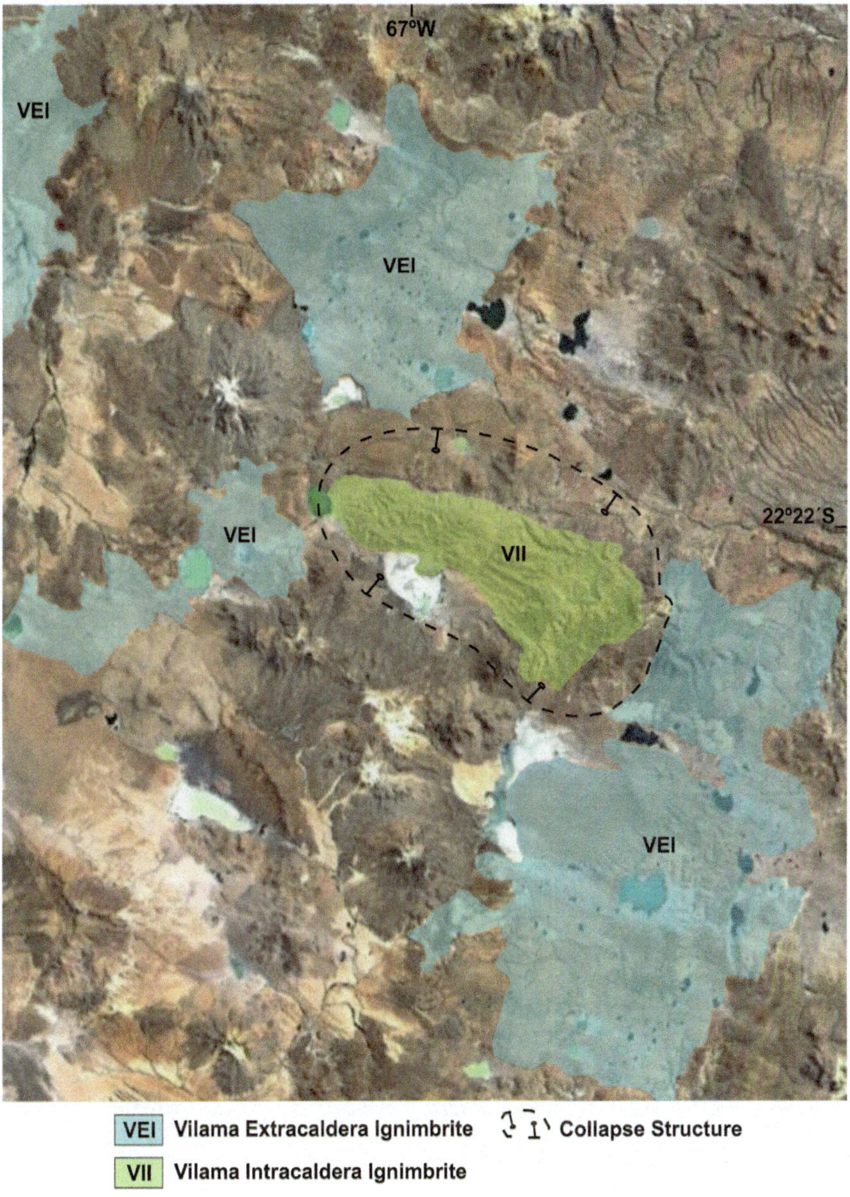

Fig. 1.6 Distribution of Vilama Ignimbrites and collapse structure

Fig. 1.7 Vilama Ignimbrite—Lower Cooling Unit (LCU): **a** Massive and unwelded ignimbrites. **b** Surge deposit. **c** Unwelded facies, high in pumice moderate in crystals, and scarce in lithics. **d** Vitroclastic textures of unwelded facies highly vesiculated pumice and moderate crystals (~20–25%) are immersed in vitric shard matrix. Plane polarized light. Upper Cooling Unit (LCU): **e** single ignimbrite sheet moderately to intensely welded. **f** Base of UCU, massive, without lithics, highly welded with development of "Fiamme" (Courtesy of M. Soler). **g** Vitrophyric section with subhorizontal jointing, well developed, usually coincident with foliation planes (Courtesy of M. Soler). **h** Upper sectors of UCU moderately rich in "fiamme" (to 10%) and lithic-poor (<1%)

Fig. 1.8 Resurgence of Vilama caldera: **a** Doming of intra-caldera ash flows. **b** intra-caldera ash flows moderately to densely welded. **c** Intra-caldera ash flows indurated and altered by vapor phase, with conspicuous jointing, in part subvertically (Courtesy of M.Soler). **d** Detail of jointing of densely welded facies, rich in crystals (Courtesy of M.Soler)

to 10%) (Fig. 1.7h) and is lithic poor (<1%) and develops an irregular and spaced jointing (Soler et al. 2007).

The intra-caldera facies outcrops are doming as a result of the resurgence of the Vilama caldera and constitute an elongated NW–SE-trending (~290°) antiform (Fig. 1.8a), which is fairly symmetrical in the NW–SE direction. The deposits consist of a thick succession of ash flows (400–700 m thick) which show important welding although variable, as well as vapor-phase alteration and individual thicknesses of 10-50 m (Fig. 1.8b). Vitrophyric portions usually occur at the base where use to show a conspicuous jointing that is sometimes subvertical (Fig. 1.8c, d) or folded, interpreted resulting from rheomorphic flow (Soler 2005).

The LCU of the Vilama Ignimbrite, given its relatively low welding and the presence of fall deposits and surges in distal facies, would be associated with the formation of eruptive columns that are well developed and are interpreted to have formed immediately before or at the beginning of the main collapse, when subsidence faults were incipient, and volcanic discharge rates and the dimension of initial vents were small. In contrast, the UCU would have been formed from dense and hot pyroclastic flows, immediately before or at the beginning of the continuous collapse of low and unstable eruptive columns, located in the boundary faults of the caldera.

An important part of those eruptive columns collapsed toward the inner side of the caldera walls. After a short time, perhaps linked to the recharge of the magma chamber, folding of the intra-caldera sequence resulted in the formation of a resurgent dome (Soler et al. 2007). Predominantly effusive post-collapse volcanism, essentially dacitic, was emplaced in the collapsed area along structures interpreted as being associated with caldera subsidence and resurgence (Fracchia 2009).

1.2.3 Coranzuli Caldera

The Coranzuli caldera, located in the back-arc of the modern Central Volcanic Zone at 23°00′S–66°15′W, is one of the most eastern Cenozoic calderas of the Northern Puna (see Fig. 1.9). It was studied in detail by Seggiaro et al. (1987, 2014b,2014b,2019) and Seggiaro (1994). This caldera is the source of four main crystal-rich dacitic ignimbrites that erupted about 6.6 Ma ago that cover an area of 1900 km^2, have a volume of 400 km^3 (dense-rock equivalent, DRE), and spread mainly north-, east-, south- and northwestwards from the caldera (see Fig. 1.9), reaching distances up to 45 km from the vent with a thickness of 40–120 m (Seggiaro et al. 2014a, b). It lies at the intersection of N-S, NW–SE, and NE-SW fault systems which controlled its emplacement (Seggiaro 1994). The Coranzulí ignimbritic succession unconformably covers Ordovician, Cretaceous, and/or Cenozoic deposits that form the basement of this region.

The syn-caldera succession is composed entirely of ignimbrites and no Plinian fallout deposits preceded the emplacement of these ignimbrites. The contacts between ignimbrite units are marked by sharp planar surfaces (Fig. 1.10), by the presence of a thin, crystal-rich, finely depleted surge layer, between two depositional units (Figs. 1.11 and 1.12), or by co-ignimbrite lag breccias. In other cases, the ignimbrites appearing as single units use to change laterally into a deposit of several flow units.

Coranzuli ignimbrites usually display columnar jointing that affects them totally or partially or more massively (Figs. 1.13 and 1.14). The co-ignimbrite lag breccias usually recognized in the proximal area are composed of abundant pumice (60 vol%), subangular fragments of dacitic lava, and minor Ordovician metapelites with a fine ash matrix (Fig. 1.15a) or are formed by dacitic lavas and Ordovician metasediment fragments (Fig. 1.15b). The lack of erosional surfaces, sedimentary deposits, or paleosoils between the ignimbrite units suggests that each flow unit was emplaced by successive pulses during a single eruptive event (Seggiaro et al. 2014a, b).

The ignimbrites of the different units are dacitic in composition, with moderate to high content in crystals (35–60 vol%) represented by plagioclase, quartz, biotite, minor sanidine, Fe Ti oxides, apatite and zircon, and poorly to densely welded, with variable pumice content (5–30%, exceptionally 50 vol%) and lithic fragment content (usually < 10 vol%; exceptionally 25%) (Fig. 1.16a, b). Pumices in all these ignimbrites have dacitic composition with variable vesicularity, subrounded to highly welded (Figs. 1.17a, b, 1.18a, b, 1.19a, b) depending on the unit or the position in the flow (see Seggiaro et al. 1987). Among the lithics are found dacitic lavas throughout the ignimbritic succession but mostly in the basal units. Ordovician metapelites and

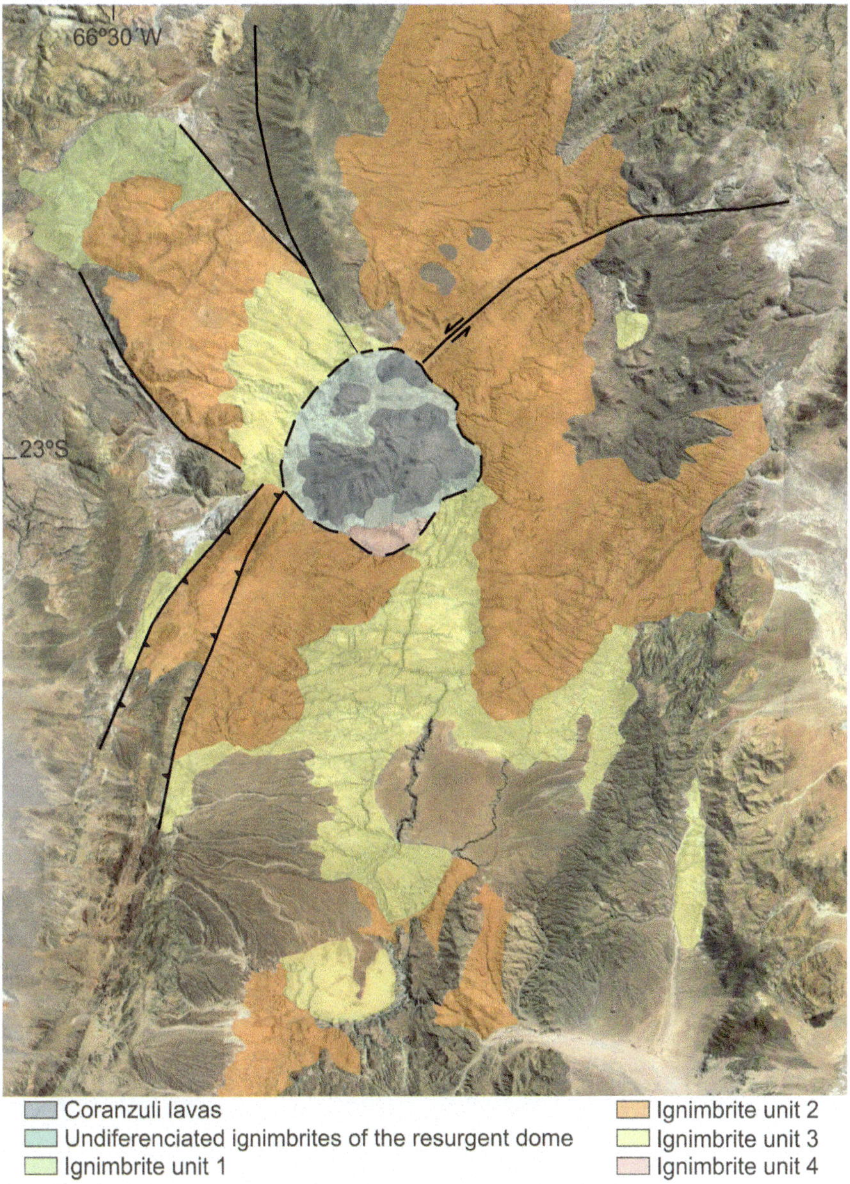

Fig. 1.9 Coranzuli caldera. with its ignimbrites and lava flows

Fig. 1.10 Sharp planar contact between two flow units of Ignibrite unit 3

Fig. 1.11 Surge layer, between two depositional units

Fig. 1.12 Detail of a finely depleted, laminated surge deposit between two flow units

Fig. 1.13 Third Ignimbrite unit with well-developed columnar jointing

Fig. 1.14 Massive ash flow of the third Ignimbrite

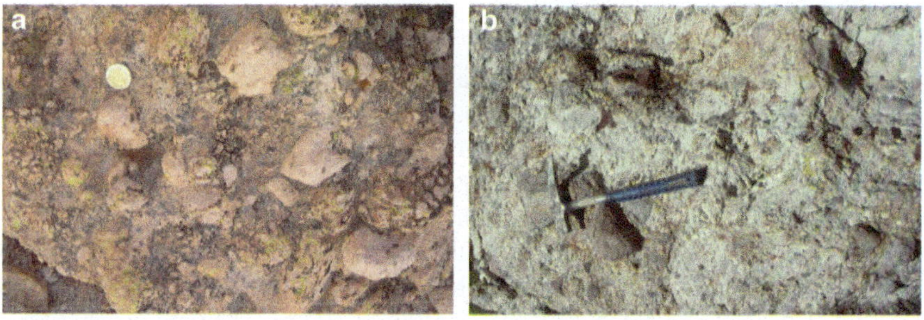

Fig. 1.15 **a** Co-ignimbrite lag breccia with abundant pumice (60 vol%), subangular fragments of dacitic lava and minor Ordovician sediments within a fine ash matrix. **b** Co-ignimbrite lag breccia with subangular fragments of dacitic lavas and minor Ordovician pelites in an abundant pyroclastic matrix

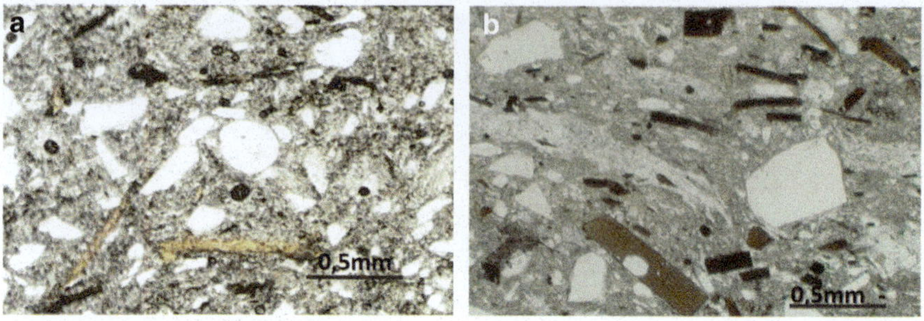

Fig. 1.16 Dacite Ignimbrites, crystal-rich (plagioclase, quartz, biotite, Fe Ti oxides) with vitroclastic texture: **a** Non-welded with vitroclastic texture and without pumice clasts. Plane polarized light. **b** Welded with alignment of stretched pumice clasts in a vitroclastic matrix. Plane polarized light

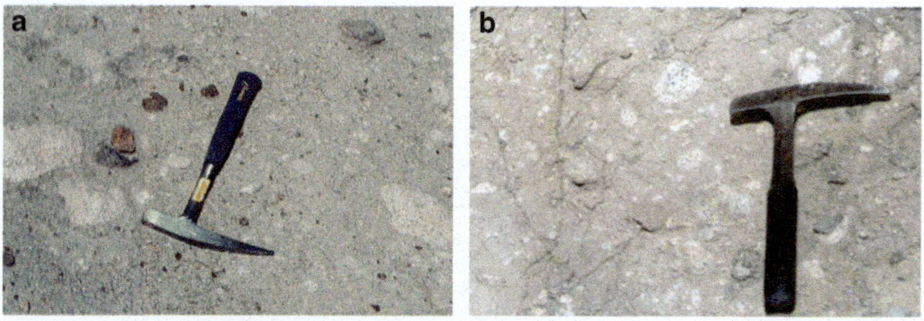

Fig. 1.17 **a** Dacitic Ignimbrite moderate in crystals and pumice, poor in lithics. **b** Dacitic Ignimbrite rich in pumice and poor in lithics

psammites are also present as lithics, mainly in the lower units; small amounts of gneisses of unexposed basement are found in the upper units and lithic fragments of dacitic ignimbrites are only observed in the uppermost unit (Seggiaro et al. 2019).

The intra-caldera ignimbritic succession was uplifted to several hundred metres by a resurgent dome affected by strong hydrothermal alteration (Fig. 1.20a). Post-caldera volcanism erupted dacitic lavas that cover the intra-caldera ignimbrites at Cerro Coranzulí (Fig. 1.20b). The lava flows in the central area of the caldera are massive-to-banded and reach thicknesses of up to 100 m. They are porphyritic or seriate, and contain phenocrysts of plagioclase, amphibole, quartz, pyroxene, and minor biotite, K-feldspar, and Fe Ti oxides (Fig. 1.21a, b).

The Coranzulí complex is emplaced at the convergence of several regional fault systems: the Coyaguayma and Ramallo NW–SE sinistral strike-slip faults, the Doncellas NE-SW dextral strike-slip fault, and the Tanque and Carahuasi N-S thrust faults with opposite vergences (Fig. 1.9). These faults allowed defining an extensional axis running in a NNE-SSW direction in the Coranzulí area. It was decisive to the formation of the caldera (Seggiaro and Hongn 1994; Seggiaro et al. 2019).

Fig. 1.18 Ignimbrite rich in pumice, poorly welded

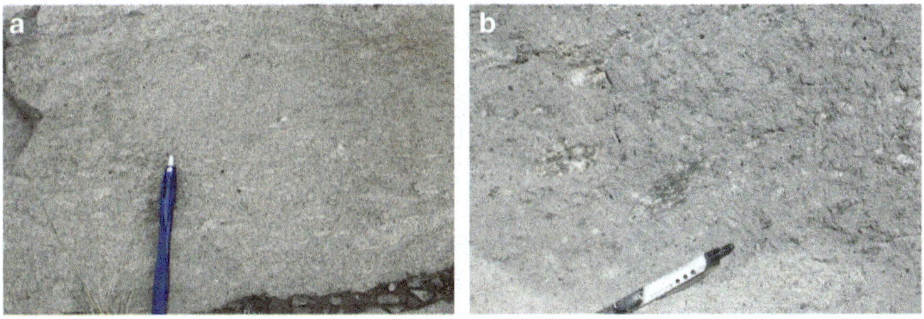

Fig. 1.19 **a** Welded ignimbrite with fiammes up to 6 cm long. **b** Densely welded facies with vitreous fiammes in a subparallel arrangement

The caldera boundaries coincide in part with the Coyaguayma fault (southern border), the Ramallo fault (northwestern border), the intersection of the Ramallo and Doncellas faults (northern border), and the Tanque fault in addition to all of their associated secondary faults (western border) (Fig. 1.9) This highlights as to how these preexisting fault systems helped define the Coranzulí caldera ring fault (Seggiaro et al. 2019).

The absence of any Plinian fallout deposits at the base of the ignimbrites was suggested by Seggiaro et al. (1987) as a result of the caldera collapsing at the beginning of the eruption, without any previous significant decompression of the magma

Fig. 1.20 **a** Intra-caldera co-ignimbritic hydrothermalized lag breccia containing blocks of a dacitic ignimbrite, and lithics of Ordovician metapelite, dacitic lava fragments. **b** The Coranzulí resurgent dome (note the hydrothermalized ignimbrites: Unit 4), at the base and dacitic lavas at the top

Fig. 1.21 **a** Coranzuli dacite porphyritic lava with phenocrysts of plagioclase, amphibole, quartz, biotite and minor pyroxene, K-feldspar, and Fe Ti oxides (hand sample showing phenocrysts 0.5–1 cm in diameter in aphanitic groundmass). **b** porphyric texture with hyalopilitic groundmass. Crossed polarized light.

chamber, suggesting a possible tectonic trigger in the formation and control of the ring faults.

The high degree of welding of some deposits, both in proximal and distal facies recognized, indicative of thermal conservation within the flows, was postulated by Seggiaro et al. (2014a, b) as fed by long-lasting eruptive pulses that produced very low eruptive columns that collapsed almost immediately (i.e., boiling over eruptions as observed in other Andean calderas, e.g., Soler et al. 2007; Guzmán and Petrinovic 2010).

The irregular distribution (i.e., non-radial) of some of the ignimbrites and the distribution and composition of their associated co-ignimbrite lag breccias led Seggiaro et al. 2019 to infer the location of the vents, suggesting that caldera collapse occurred with different ring fault sectors operating at different times. They postulated a “trapdoor” behavior during the first pulse of the caldera collapse considering the only north-to-northwestwards distribution of the first ignimbrite. In the case of the second ignimbrite flow units, they suggested a more continuous subsidence of the caldera block (i.e., piston-like) considering its radial emplacement. The last ignimbrite, associated with a co-ignimbrite lag breccia containing lithic clast fragments of ignimbrites, in addition to the same lithic fragments observed in the lower ignimbrite units, was considered as a result of a widening of the caldera rim and of a change in the collapse dynamics, this time behaving again as a "trapdoor". An average age of 6.6 ± 0.15 Ma represents the age of the caldera eruption and the K–Ar biotite age of 6.73 ± 0.2 Ma for the rhyodacites to the south of Cerro Coranzulí the post-caldera lavas emission (Seggiaro et al. 2014a, b). Aquater (1979) suggests that the emplacement of these lavas occurred shortly after the formation of the caldera.

1.3 Southern Puna Calderas

Widespread volcanism took place in Southern Puna with ignimbrite activity at 11–10 Ma succeeded by andesitic lavas and dome complexes at composite volcanoes (e.g., Antofalla volcano). Compressive deformation continued. A major change in volcanism and deformation occurred at about 6.7–6 Ma as mafic lavas associated with normal and strike-slip faults began erupting, succeeded by initial eruptions in the back-arc of the voluminous ignimbrites of the Cerro Galan caldera (Merihuaca 6.4–5.14 Ma, Real Grande 5.1–4.2 Ma) and in the arc of Laguna Verde–Laguna Amarga–Laguna Vallecito ignimbrites. The major eruption of Cerro Galan occurred at 2.1 Ma followed by near intraplate mafic lavas and more arc-like elsewhere. The eruptions of Cerro Blanco caldera and associated mafic rocks occurred at 1 Ma. The volcanic arc front stabilizes in the region of modern CVZ.

The largest ignimbrite in the Southern Puna is (Figs. 1.22 and 1.23) the ~6–2 Ma Cerro Galán caldera (~1200 km^3); other important ignimbrites are the 4.8–3.8 Ma Laguna Amarga–Laguna Vallecito (~70 km^3) ignimbrites and those associated with the < 1 Ma Cerro Blanco caldera (~150 km^3).

1.3.1 Cerro Galan Caldera

The Cerro Galán caldera (26° 00′S, 66° 55′W, Fig. 1.24), with nine extra-caldera ignimbrites associated, recording a continuous volcanic activity from 6 Ma to ~2 Ma.

The initial activity was recorded in the Toconquis Group (Sparks et al. 1985), see (Fig. 1.25), divided into the 6.4–5.5 Ma Blanco/Merihuaca, 4.8 Ma Pitas, 4.7 Ma Real Grande, 4.5 Ma Vega, and 3.8 Ma Cueva Negra Ignimbrites (Folkes et al. 2011a). The Blanco Ignimbrite is massive, and relatively pumice- (15%) and crystal-rich (20%) with a large ash fraction.

The overlying Merihuaca Ignimbrites (Lower, Middle, and Upper members) are massive, pumice-rich (up to 40%), lithic-poor (<5%), and crystal-rich (25–30%) and reach a maximum thickness of ~45 m (Fig. 1.26a–d). These ignimbrites are succeeded by Pitas Ignimbrite, massive, pumice-rich (30%), lithic-poor(<5%) with pumice-concentrated zones, and a maximum thickness of ~60 m. The base of the

Fig. 1.22 Satellite images from Southern Puna calderas

overlying Real Grande Ignimbrite is marked by a surge and ashfall basal layer that is a good marker horizon (Fig. 1.26e, f) and indicates a time break between the Pitas and Real Grande Ignimbrites. This last ignimbrite is the thickest in the Toconquis Group (~120 m at its thickest points). It is massive, pumice-rich (up to 40%), crystal-rich (25–30%), lithic-poor (<5%) (Fig. 1.27f), and contains conspicuous zones of pumice concentration with pumice clasts up to 1 m in diameter and are lithic-rich in scarce levels (Fig. 1.27g). Immediately above the Real Grande Ignimbrite is the thin (2–3 m) pumice- (50%) and lithic-rich (25%) Vega Ignimbrite (Folkes et al. 2011a).

As a synthesis, it can be said that all ignimbrites belonging to the Toconquis Group possess similar features. They are all crystal-rich (25–35%), pumice-rich (20–40%), and lithic-poor (<5%). They are entirely non-welded, but indurated, and commonly contain 1–4 m-thick pumice concentration zones and in general contain surge and ashfall deposits as basal levels < 1 m thick, succeeded by massive deposits. The Toconquis pumices and whole rock ignimbrites consist of 78–90% matrix with plagioclase and biotite, smaller amounts of quartz and Fe Ti oxides, and little to no sanidine. The Real Grande Ignimbrite can also contain minor amounts of hornblende

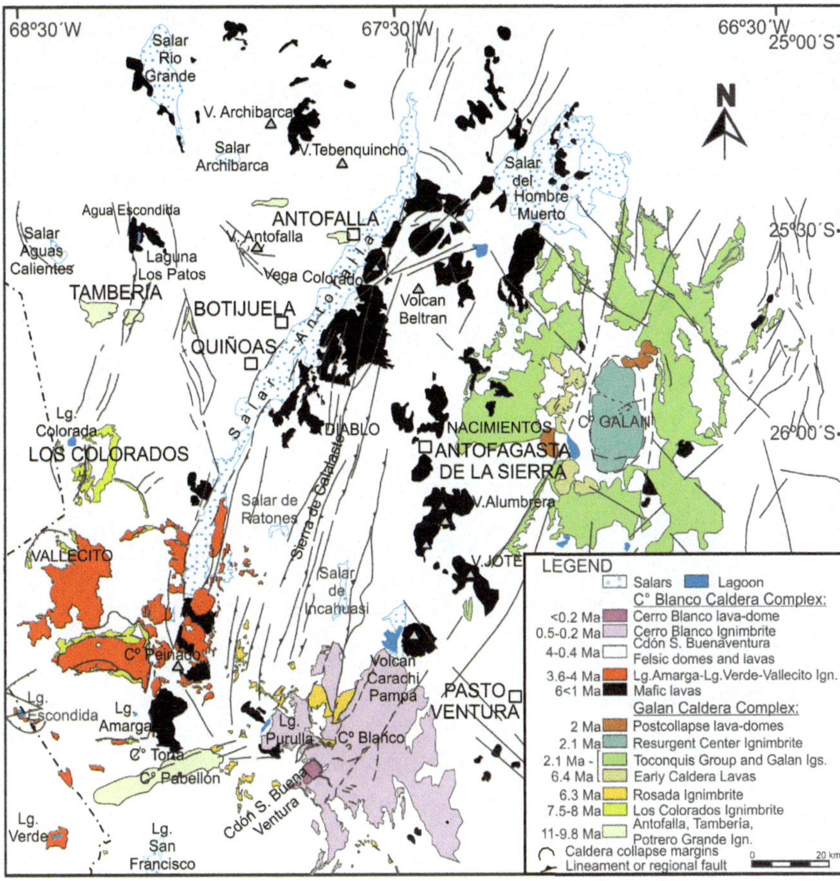

Fig. 1.23 Spatial and temporal distribution of Southern Puna Calderas (modified from Kay et al. 2010)

and orthopyroxene. The Cueva Negra Ignimbrite is distinguished from the Toconquis Group Ignimbrites because a period of quiescence existed between them (Folkes et al. 2011a). This Ignimbrite is found only on the eastern flanks of the present caldera margin where it reaches the maximum thickness of ~80 m. It is characterized by a welded texture throughout most of its stratigraphic thickness. Compositionally, it is crystal rich (25–30%), pumice rich (15–25%), and lithic poor (<5%). The Toconquis Group (GT) and the Cueva Negra Ignimbrite extend mainly toward the W and also to the S and N of the current caldera, with an approximate extension of 70 km and with a maximum thickness of 240 m. Its volume was estimated at 650 km^3 (Folkes et al. 2011a). The youngest large-volume ignimbrite is the Cerro Galán Ignimbrite (2.06–2.13 Ma, Kay et al. 2011). This massive ignimbrite throughout most of its extent (Fig. 1.28a) is extremely crystal-rich (35–50%), pumice-poor (<10%), and lithic-

Fig. 1.24 Cerro Galan caldera and associated volcanic rocks

poor (<2%) (Fig. 1.28c, f). It commonly fills preexisting topography as valley-ponded and extends up to 100 km from the edge of the caldera (Fig. 1.28b).

The outflow facies reported to average 51% matrix, 11% quartz, 18% plagioclase, 6% sanidine, and 7% biotite, and the intra-caldera deposits 58% matrix, 9% quartz, 17% plagioclase, 8% sanidine, and 7% biotite; they also contain titanomagnetite and trace amounts of zircon, apatite, and allanite (Francis et al. 1989). Localized pumice-rich levels are observed in medial to distal localities of the outflow sheet (Fig. 1.28d). The western outflow sheet is thin in proximal areas (~20 m) and thickens to ~40 m at distances >40 km from the caldera structural margin (Sparks et al. 1985). Proximal lithic breccias and ashfall deposits have not been recognized (Cas et al. 2011). Sections with columnar jointing possibly controlled by water table are recognized in distal facies (Fig. 1.29b).

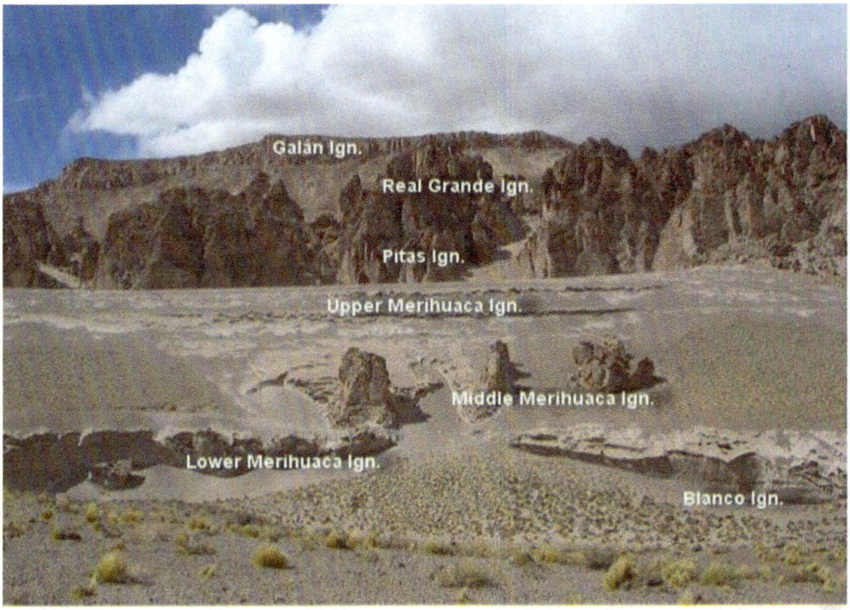

Fig. 1.25 Toconquis Group Ignimbrites capped by the Cerro Galán Ignimbrite at Río de Las Pitas, ~12 km west of the present structural margin

The Cerro Galán Ignimbrite was interpreted as an ignimbrite of LARI type (Low Aspect Ratio Ignimbrite: 4.5×10^{-4}; Cas et al. 2011) that flowed at low speed and only in distal areas it had the ability to overcome the topography, but without showing stratified deposits of type "veneer" (Cas et al. 2011).

The Cerro Galán caldera has a structural diameter of 27 × 16 km elongated in WNW-ESE direction (Folkes et al. 2011a) and a diameter of 35 × 20 km elongated in N-S direction (Sparks et al. 1985)). It contains a central resurgent dome that reaches 1.4 km in height (Sparks et al. 1985) integrated predominantly by the Cerro Galán Ignimbrite (Fig. 1.24) although at its southern end it contains portions of the basement (Folkes et al. 2011a). The Cerro Galán caldera has been controlled by two faults of N-S orientation affecting the basement (Francis et al. 1989; Sparks et al. 1985). However, Folkes et al. 2011a considered that there is no evidence of structural control of the east edge.

The style of the collapse of the caldera was interpreted by Sparks et al. (1985) as a piston-like style and reinterpreted by Folkes et al. (2011a), who proposed a "piecemeal" style or "trapdoor" type, with the W edge being sunken. The depth of the collapse was at least 1.4 km—the equivalent to the thickness of the intra-caldera ignimbrite (Folkes et al. 2011a). The collapse of the caldera was associated with the Cerro Galán Ignimbrite with an estimated volume of 630 km^3 (Cas et al. 2011; Folkes et al. 2011a). The preserved area extra-caldera, that occupies the Cerro Galán Ignimbrite is 2,400 km^2 (Folkes et al. 2011a), is distributed radially around

Fig. 1.26 Merihuaca Ignimbrite (MI), **a** Massive, non-indurated, pumice rich. **b** Pumice-rich (up to 40%), lithic-poor (<5%), crystal-rich. **c** Middle facies (MI) pumice rich, non-indurated. **d** Proximal facies lithic clast-rich. **e** Pitas Ignimbrite (at the base), massive, pumice-rich (30%), lithic-poor (<5%) with pumice concentrated zones. On top, Real Grande Ignimbrite, at its base a thin-bedded surge deposit (**f**)

the caldera, reaching distances of 80 km from its structural edge, and has a maximum thickness extra-caldera of 80 m (Cas et al. 2011). The eruption associated with the caldera formation was not preceded by a Plinian eruption and was of type "boiling over", "fontana style" (Sparks et al. 1985; Cas et al. 2011; Folkes et al. 2011a), generating a pyroclastic current with high discharge rate, constant and sustained, without the development of a significant eruptive column (Cas et al. 2011).

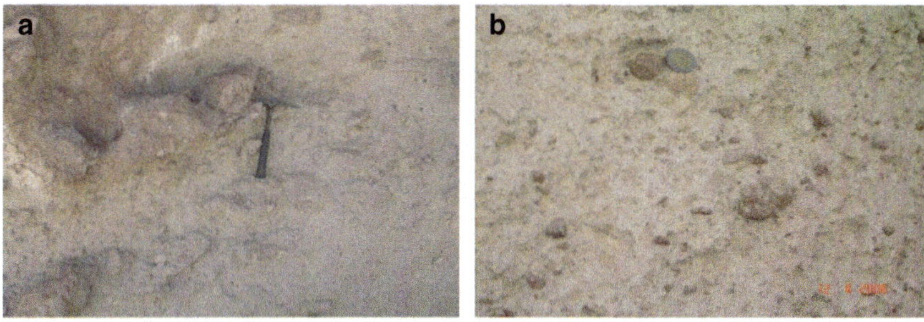

Fig. 1.27 Real Grande Ignimbrite: **a** Pumice-rich (up to 40%), crystal-rich (25–30%), lithic-poor (<5%). **b** Scarce levels poorly sorted, rich in lithics

Fig. 1.28 Galán Ignimbrite (GI). **a** Massive, indurated, absence of basal Plinian deposits. **b** Distal pyroclastic flow that extends up to 100 km from the edge of the caldera. **c** GI extremely crystal-rich (35–50%), pumice-poor (<10%), and lithic-poor (<2%). **d** Pumice-rich level. **e** Poor in pumice and moderate in lithic clasts and crystals. **f** Resurgent doming of intra-caldera ignimbrite

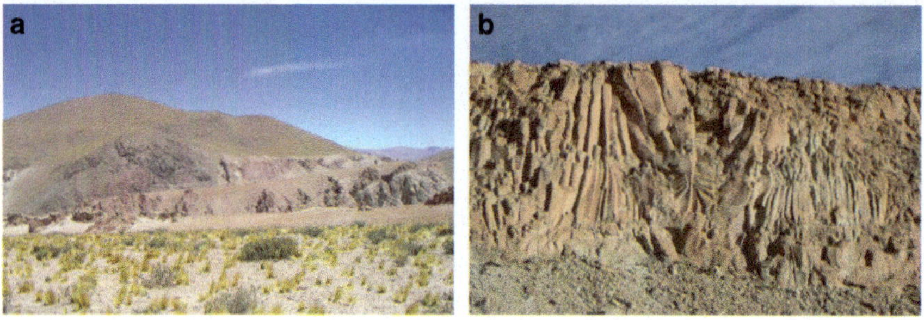

Fig. 1.29 **a** Post-caldera extrusion along its margin. **b** Granada Ignimbrite: distal facies with columnar jointing possibly controlled by water table

The post-caldera activity is on the northern edge of the resurgent dome and in N and E areas of the current caldera. It consists of massive or stratified pyroclastic ash flows, lava domes (Fig. 1.29a), and block and ash deposits whose post-data the Cerro Galán Ignimbrite activity in at least 0.5 Ma (Folkes et al. 2011a).

1.3.2 Cerro Blanco Volcanic Complex

Cerro Blanco Volcanic Complex (CVCB), located in the southern limit of Puna (~26° 46′S, 67° 44′W), corresponds to a Middle Pleistocene–Holocene system of nested calderas, with associated domes and pyroclastic deposits of rhyolitic/rhyodacitic composition (Fig. 1.30). The CVCB has been the subject of numerous detailed studies in the last decade (e.g., Arnosio et al. 2005, 2008; Viramonte et al. 2008; Montero López 2009; Montero López et al. 2010, Fernández-Turiel et al. 2013; Báez 2014; Báez et al. 2015). In the CVCB, a nested caldera system (Caldera del Cerro Blanco, Caldera de Robledo, and Caldera Pie de San Buenaventura; Fig. 1.30) has been recognized based on a series of morphostructural features (Seggiaro et al. 2000; Arnosio et al. 2005; Viramonte et al. 2005a, b). Subsequently, as a result of the gravimetric data obtained in that area by Di Filippo et al (2008), the deformation patterns defined in it by INSAR by Brunori et al. (2013), and the detail studies by Báez et al. (2015), a single caldera has been proposed instead of the nested caldera system. This Cerro Blanco caldera, of redefined elliptical morphology (~13 × 10 km, Fig. 1.31), is formed by an internal depression limited by thrust faults and an incipient external depression limited by normal faults, as a result of the compression/extension regimen that characterized the region during its evolution (Báez et al. 2015).

In the evolution of the CVCB were recognized (Báez et al. 2015) three main eruptive epochs (Fig. 1.30). (a) the Cortaderas eruptive epoch (Middle Pleistocene) represented by explosive boiling over eruptions with the generation of huge volumes of pyroclastic density currents (Barranca Blanca and Carachi Ignimbrite) from an

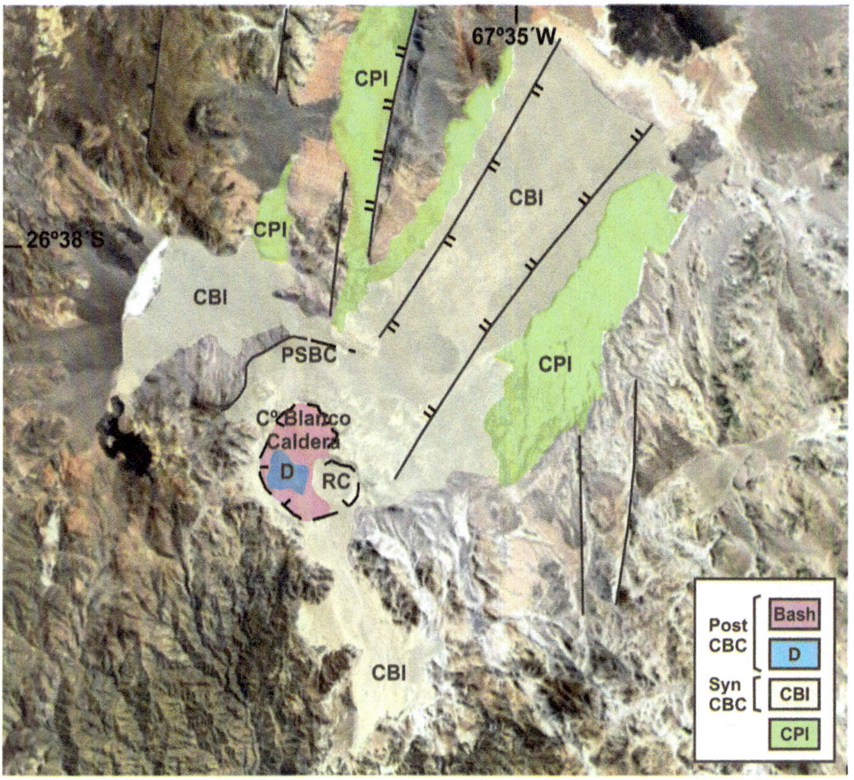

Fig. 1.30 Cerro Blanco Volcanic Complex: CBC: Cerro Blanco caldera; RC: Robledo caldera; PSBC: Pie de San Buenaventura caldera; CPI (Campo Piedra Pómez Ignimbrite); Syn Cerro Blanco caldera CBC(CBI, C Blanco Ign.); Post-CBC (D-domes; Bash:Block and ash deposits)

unknown source. (b) The Campo de la Piedra Pómez eruptive epoch (Upper Pleistocene), formed by at least one boiling over, explosive eruption characterized by the generation of huge volumes of pyroclastic density currents (Campo Piedra Pómez Ignimbrite) possibly associated with a volcano-tectonic depression. (c) The Cerro Blanco eruptive epoch (Holocene), which includes the pre-, syn-, and post-Cerro Blanco caldera episodes. The pre-caldera activity was effusive and was represented by block and ash flow deposits. The syn-caldera activity was explosive with a Plinian/sub-Plinian eruptive style with simultaneous pyroclastic density current generation, represented by Cerro Blanco Ignimbrite and associated pyroclastic fall deposits. The post-caldera activity was effusive/explosive and is represented by post-caldera domes and associated block and ash and Vulcanian fall deposits. This eruptive epoch constitutes one of the greatest Holocene volcanic events in the Central Andes.

The Barranca Blanca Ignimbrite and the Carachi Ignimbrite, representative of the Cortaderas eruptive epoch, are rhyodacitic, massive, not welded to moderately welded, and outcrop very punctually in the surroundings of the lagoon Carachipampa.

Fig. 1.31 Cerro Blanco caldera—D domes

The lava of the Carachipampa volcano (750 ka, Risse et al. 2008) covers quaternary terraces that include these ignimbrites indicating the minimum age of them.

The Campo de la Piedra Pómez Ignimbrite (CPI), a product of the second eruptive epoch, emerges to the north from the caldera of Cerro Blanco (Fig. 1.30), occupying an area of ~250 km^2 with a minimum total volume of ~17 km^3 and DRE volume of ~9 km^3 (Báez et al. 2015). The CPI has varying percentages of pumice (~15 a 50 vol. %), of lithic fragments (~2–50 vol.%), and in general is poor in crystals (~10–30 vol.%). The dominant pumice are white and rhyolitic, and the gray to black pumices, less abundant, are traquiandesitic.

These variations in the percentages of pumice, lithics, and crystals respond to variations of internal facies. This Ignimbrite is characterized for being formed by poorly selected mass facies (Fig. 1.32). However, in proximal areas it is partially

Fig. 1.32 Massive CPI characterized by the absence of ashfall deposits

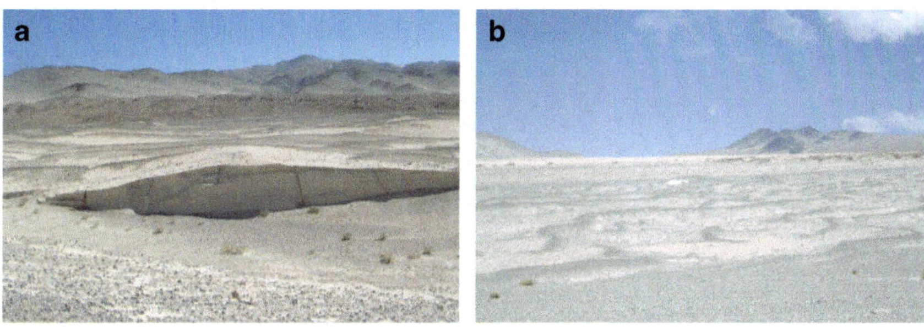

Fig. 1.33 **a** CPI outcrops in the volcano-tectonic depression of Carachipampa. **b** Detail in that sector of Yardang field developed on CPI

welded and develops lithic-rich facies. In distal areas, it has a significant vertical and lateral variation of facies (Fig. 1.33a, b). A distinctive feature of this Ignimbrite is the absence of ashfall deposits and strong topographic control.

Although there are numerous radiometric dates of the CPI, Báez et al. (2015) adopt an age for the CPI of ~73 ka because it is the age obtained with the more precise method (40Ar/30Ar in sanidine; Hora et al. 2010).

The absence of a well-defined caldera structure to which can be associated the CPI and the outcrops of lithic breccias within the CPI, that match with the trace of some structures of the volcano-tectonic depression of Carachipampa, has led Báez et al. (2015) to propose an alternative model where the volcanic structure that has generated the CPI corresponds with that volcano-tectonic depression.

The pre-caldera Cerro Blanco records correspond to block and ash deposits with blocks of rhyolite lavas in an ash matrix. They crop up in the walls of the caldera and unconformably cover andesitic/dacitic lavas and Miocene volcaniclastic deposits, belonging to the Volcanic Complex La Hoyada. The syn-caldera activity was represented by the Cerro Blanco Ignimbrite (CBI) and pyroclastic fall deposits. They have been previously interpreted as a result of the collapse of the Cerro Blanco caldera (Arnosio et al. 2005). The outcrops of the CBI are distributed to the west, to the south, and to the north of the caldera of Cerro Blanco (Fig. 1.30) occupying an area of ~527 km^2. A minimum volume was calculated for it of ~17 km^3 and DRE volume of ~9 km^3 (Báez et al. 2015). The CBI is a white rhyolitic, non-welded ignimbritic deposit and without vapor-phase crystallization (Fig. 1.34), formed by varying percentages of lapilli-size fragments/block of rhyolitic pumice (~20–30 vol.%) and lithic fragments (~20–30 vol.%) in a matrix of fine ash size (Fig. 1.35). Stratified facies of veneer type on the topographic highs and the associated pyroclastic fall deposits are features that differentiate the CBI from the CPI (Báez et al. 2015).

The Plinian fall deposits associated with the CBI are relatively well selected with lapilli to block size (Fig. 1.36) and variable thicknesses between 0.20 m and 1.2 m.

Fig. 1.34 CBI massive and non-welded

Fig. 1.35 CBI with Plinian deposit interspersed

Fig. 1.36 CBI with rhyolitic pumice and lithics in ash size matrix non-welded

Considering the ages obtained by method C14 for the CBI as the most adjusted one, the post-boiler volcanism would have an age of ~4000–5500 BC (Holocene).

A series of overlapping rhyolitic domes located on the southwest topographic edge of the Cerro Blanco caldera (Figs. 1.31 and 1.37) correspond to the post-caldera volcanism. Inside the Caldera, block and ash deposits and volcaniclastic deposits associated with phreatic/phreatomagmatic explosion craters are recognized as post-caldera records.

The Cerro Blanco caldera superficial morphology can be interpreted as a result of an advanced state of subsidence (sensu Acoccella 2007). Available gravimetry data (Di Filippo et al. 2008) suggest a block structure (piecemeal) for the Cerro Blanco caldera, with subsidence differential for each of these and a minimum subsidence average of >700 m (Báez et al. 2015).

Fig. 1.37 Rhyolithic dome inside the Cerro Blanco caldera

References

Acocella V (2007) Understanding caldera structure and development: an overview of logue models compared to natural calderas. Earth-Sci Rev 85(3):125–160

Aquater (1979). Estudio del potencial geotérmico de la provincia de Jujuy, República Argentina, Fase de reconocimiento. Secretaría de Estado de Minería: 129 pág. Informe inédito. Buenos Aires

Arnosio M, Becchio R, Viramonte JG, Groppelli G, Norini G, Corazzato C (2005) Geologia del Complejo Volcanico Cerro Blanco (26° 45′ LS- 67° 45′ LO), Puna Austral, *en* 16o Congreso Geologico Argentino: La Plata Buenos Aires. Asociacion Geologica Argentina 1:851–858

Arnosio M, Becchio R, Viramonte JG, de Silva S, Viramonte JM (2008) Geocronologia e isotopia del Complejo Volcanico Cerro Blanco: un sistema de calderas cuaternario (73–12 ka) en los Andes Centrales del sur (resumen), *en* 17° Congreso Geologico Argentino Jujuy. Argentina: Asociacion Geologica Argentina 1:177–178

Báez W (2014) Estratigrafía volcánica, estilos eruptivos y evolución del Complejo Volcánico Cerro Blanco. Salta, Argentina, Universidad Nacional de Salta, tesis doctoral, Puna Austral, p 204

Báez W, Arnosio M, Chiodi A, Yáñez-Ortiz A, Viramonte JG, Bustos E, Guido G, López F (2015) Estratigrafía y evolución del Complejo Volcánico Cerro Blanco, PunaAustral, Argentina. Revista Mexicana de Ciencias Geológicas 31(1):29–49

Beck S, Zandt G (2002) The nature of orogenic crust in the central Andes. J Geophys Res 107, 2230. https://doi.org/10.1029/2000JB000124

Brunori CA, Bignami C, Stramondo S, Bustos E (2013) 20 years of active deformation on volcano caldera: joint analysis of InSAR and AInSAR techniques. Int J Appl Earth Obs Geoinf 23:279–287

Cas RAF,Wright HMN, Folkes CB, Lesti C, Porreca M, Giordano G, Viramonte JG (2011) The flow dynamics of an extremely large volume pyroclastic flow, the 2.08-MaCerro Galán Ignimbrite, NW Argentina, and comparison with other flow types. Bull Volcanol 73(10):1583–1609. https://doi.org/10.1007/s00445-011-0564-y

Coira B, Malhburg Kay S, Viramonte J (1993) Upper Cenozoic magmatism of the Argentina Puna. Int Geol Rev 35(8):677–720

Coira BL, Caffe P, Kay SM, Diaz A, Ramirez S (1996) Complejo volcánico Vilama —sistema caldérico del Cenozoico superior en Puna Jujuy. XIII Congreso GeológicoArgentino Actas 3, Buenos Aires, Argentina 603–620

Coira B, Caffe PJ, Ramírez A, Chayle W, Díaz A, Rosas SA, Pérez A, Pérez EMB, Orosco O, Martínez M (2004) Hoja Geológica 2366-I Mina Pirquitas 1:250.000.Instituto de Geología y Recursos Minerales, Servicio Geológico Minero Argentino, Buenos Aires Boletín 269

Cheilletz A, Clark AH, Farrar E, Arroyo Pauca G, Pichavant M, Sandeman HA (1992) Volcano-stratigraphy and 40Ar/39Ar geochronology of the Macusani ignimbrite field: monitor of the Miocene geodynamic evolution of the Andes of southeast Peru. Tectonophysics 205:305–327

Christiansen RL (1979) Cooling units and composite sheets in relation to caldera structure. Spec Pap Geol Soc Am 180:29–42

de Silva SL (1989) Altiplano-Puna volcanic complex of the central Andes. Geology 17:1102–1106. https://doi.org/10.1130/0091-7613(1989)017b1102:APVCOTN2.3.CO;2

de Silva S, Gosnold WD (2007) Episodic construction of batholiths: insights from the spatiotemporal development of an ignimbrite flare-up. J Volcanol Geoth Res 167:320–325

de Silva SL, Zandt G, Trumbull RB, Viramonte JG, Salas G, Jimenez M (2006) Large-scale silicic volcanism in the Central Andes — a tectonomagmatic perspective. In: Troise C, de Natale G, Kilburn CRJ (eds) Mechanisms of activity and unrest at large calderas. Geological Society of London SpecialPublication, 269, pp 47–63

Di Filippo M, Di Nezza M, Colombi A, Viramonte JG, Toro B (2008) Estructura gravimetrica del Complejo Volcanico Cerro Blanco, Puna Austral Argentina. 17° Congreso Geologico Argentino, Actas 1, Jujuy, Argentina, pp 203–204

Fernández-Turiel JL, Saavedra J, Perez-Torrado JF, Rodriguez-Gonzalez A, Carracedo JC, Osterrieth M, Carrizo JI, Esteban G (2013) The largest holocene eruption of the Central Andes found, *en* American Geophysical union fall meeting. American Geophysical Union, San Francisco, California, pp V13D-2639

Folkes CB, Wright HM, Cas RAF, de Silva SL, Lesti C, Viramonte JG (2011a) A reappraisal of the stratigraphy and volcanology of the Cerro Galán volcanic system,NW Argentina. In: Cas RAF, Cashman K (eds) the Cerro Galan Ignimbrite and Caldera: characteristics and origins of a very large volume ignimbrite and its magma system. Bull Volcanol 73(10):1427–1454. https://doi.org/10.1007/s00445-011-0459

Folkes CB, de Silva SL, Wright HM, Cas RAF (2011b) Geochemical homogeneity of a long-lived, large silicic system; evidence from the Cerro Galán caldera, NW Argentina. Bull Volcanol 73(10):1455–1486

Fracchia D (2009) Volcanismo postcolapso de la Caldera Vilama, Mioceno superior, Puna argentino–boliviana: Mecanismos eruptivos y petrogénesis. Tesis doctoral Fac. de Ciencias Nat. y Museo, Universidad Nacional de La Plata (inédita), 459 p., La Plata

Francis PW, Sparks RSJ, Hawkesworth CJ, Thorpe RS, Pyle DM, Tait SR, Mantovani MS, McDermott R (1989) Petrology and geochemistry of volcanic rocks of the Cerro Galán caldera, northwest Argentina. Geol Mag 126:515–547

Gardeweg M, Ramírez C (1987) La Pacana caldera and the Atana Ignimbrite—a major ash-flow and resurgent caldera complex in the Andes of northern Chile. Bull Volcanol 49:547–566

Guzmán S (2009) Petrología y relaciones tectono-magmáticas del Complejo Volcánico Pucarilla – Cerro Tipillas, Provincia de Salta. Tesis doctoral (inédita), Universidad Nacional de Salta, 220 p., Salta

Guzmán S, Petrinovic I (2010) The Luingo caldera: the south-easternmost collapse caldera in the Altiplano-Puna plateau, NW Argentina. J Volcanol Geotherm Res 194:174–188. https://doi.org/10.1016/j.jvolgeores.2010.05.009

Hildreth W (1981) Gradients in silicic magma chambers: implications for lithospheric magmatism. J Geophys Res 86:10153–10192

Hora JM, Singer BS, Jicha BR, Beard BL, Johnson CM, de Silva S, Salisbury M (2010) Volcanic biotite-sanidine 40Ar/39Ar age discordances reflect Ar partitioning and pre-eruption closure in biotite. Geology 38(10):923–926

Kay SM, Coira B (2009) Shallowing and steepening subduction zones, continental lithosphere loss, magmatism and crustal flow under the central Andean Altiplano– Puna plateau. In: Kay, S.M., Ramos, V.A., Dickinson, W.M. (Eds.), Backbone of the Americas: Shallow Subduction, Plateau and Ridge and Terrane Collisions: Geological Society America Memoir, 204, pp. 229–260

Kay SM, Coira B, Viramonte J (1994) Young mafic back arc volcanic rocks as indicators of continental lithospheric delamination beneath the Argentine Puna Plateau, central Andes. J Geophys Res [Solid Earth] 99 (B12):24323–24339

Kay SM, Mpodozis C, Coira B (1999) Magmatism, tectonism and mineral deposits of the Central Andes (22°–33°S latitude). In: Skinner BJ (ed) Geology and ore deposits of the central Andes: Society of Economic Geology Special Publication 7:27–59

Kay SM, Coira BL, Mpodozis C (2008) Field trip guide to the Neogene to Recent evolution of the Puna plateau and the southern Central Volcanic Zone. In: Kay, S.M., Ramos, V.A. (Eds.), Field trip Guides to the Backbone of the America in the Southern and Central Andes: Geological Society America Field Guide Series, 13, pp. 117–181

Kay S, Coira B, Caffe P, Chen C (2010) Regional chemical diversity, crustal and mantle sources and evolution of Central Andean Puna plateau ignimbrites. J Volcanol Geotherm Res 198: 81–110

Kay SM, Coira B, Wörner G, Kay RW, Singer S (2011) Geochemical, isotopic and single crystal 40Ar/39Ar age constraints on the evolution of the Cerro Galán Ignimbrites. Bull Volcanol. https://doi.org/10.1007/s00445-010-0410-7

Lindsay JM, de Silva S, Trumbull R, Emmermann R, Wemmer K (2001) La Pacana caldera, N. Chile: a re-evaluation of the stratigraphy and volcanology of one of the world's largest resurgent caldera. J Volcanol Geotherm Res 106:145–173. https://doi.org/10.1016/S0377-0273(00)00270-5

Martí J, Geyer A, Folch A (2009) A genetic classification of collapse calderas based on field studies, and analogue and theoretical modelling. In: Thordarson T, Self S, Larsen G, Rowland SK,

Hoskuldsson A (eds) Studies in volcanology: the legacy of George Walker. Special Publications of IAVCEI, 2. Geological Society, London, pp 249–266
McGlashan N, Brown LD, Kay SM (2008) Crustal thicknesses in the Central Andes from teleseismically recorded depth phase precursors. Geophys J Int 175:1013–1022
Montero-Lopez MC (2009) Estructura y magmatismo neogeno-cuaternario en la sierra de San Buenaventura (Catamarca): su vinculacion con la terminacion austral de la Puna: Salta. Universidad Nacional de Salta, tesis doctoral, Argentina, p 255
Montero-Lopez MC, Hongn F, Brod JA, Seggiaro R, Marrett R, Sudo M (2010) Magmatismo acido del Mioceno Superior-Cuaternario en el area de Cerro Blanco-La Hoyada. Puna Sur: Revista De La Asociacion Geologica Argentina 67(3):329–348
Morgan VI GB, London D, Luedke RG (1998) Petrochemistry of late Miocene peraluminous silicic volcanic rocks from the Morococala field, Bolivia. J Petrol 39:601–632
Ort M (1993) Eruptive processes and caldera formation in a nested downsag collapse caldera: Cerro Panizos, Central Andes Mountains. J Volcanol Geotherm Res 56:221–252. https://doi.org/10.1016/0377-0273(93)90018-M
Ort M, Coira B, Mazzoni M (1996) Generation of a crust–mantle magma mixture: magma sources and contamination at Cerro Panizos, central Andes. Contrib Miner Petrol 123:308–322
Pacheco J, Ramírez V (1997) Hoja Geológica Cañapa/Alota 5929/6029: Servicio Nacional de Geológica y Mineria Carta Geológica de Bolivia Publicacíon SGM Serie I–CGB-45, scale 1:100,000, 1 sheet
Petrinovic IA, Mitjavilla J, Viramonte JG, Marti J, Becchio R, Arnosio M, Colombo F (1999) Geoquímica y geocronología de las secuencias neógenas de trasarco, en el extremo oriental de la cadena volcánica transversal del Quevar, noroeste de Argentina, Acta Geológica Hispana, 34:255–273
Petrinovic IA, Martí J, Aguirre-Díaz GJ, Guzmán SR, Geyer A, Salado Paz N (2010) The Cerro Aguas Calientes caldera, NW Argentina: an example of a tectonically controlled polygenetic collapse caldera, and its regional significance. J Volcanol Geotherm Res 194:15–26. https://doi.org/10.1016/j.jvolgeores.2010.04.012
Risse A, Trumbull RB, Coira B, Kay SM, Bogaard PVD, (2008) 40Ar/39Ar geochronology of mafic volcanism in the back-arc region of the southern Puna Plateau, Argentina. J South Am Earth Sci 26(1):1–15
Schmitt AK, de Silva SL, Trumbull R, Emmermann R (2001) Magma evolution in the Purico ignimbrite complex, northern Chile: evidence for zoning of dacitic magma by injection of rhyolitic melts following mafic recharge. Contrib Mineral Petrol 140:680–700
Schneider A (1987) Eruptive processes, mineralization and isotopic evolution of the Los Frailes, Kari Kari Region Bolivia. Revista Geológica de Chile 30:27–33
Seggiaro RE (1994) Petrología, Geoquímica y Mecanismos de Erupción del Complejo Volcánico Coranzulí. Ph.D. Thesis. Universidad Nacional de Salta, Argentina
Seggiaro R, Hongn F (1994) Tectónica Transcurrente Asociada al Volcán Coranzulí, Jujuy,Argentina. vol. 1. 7° Congreso Geológico Chileno, Concepción, Chile, pp 169–173
Seggiaro RE, Gorustovich SA, Martí J (1987) Las ignimbritas del Complejo Volcánico Coranzulí (Puna Argentina-Andes Centrales) Estudios. Geológicos 43:345–358
Seggiaro R, Hongn F, Folguera A, Clavero J (1999) Hoja Geológica 2769 – II, Paso de San Francisco, 1:250.000: Buenos Aires, Servicio Geológico Minero Argentino, Programa Nacional de Cartas Geológicas, 1 Mapa
Seggiaro R, Guzmán S, Martí J, Montero-López C, López E (2014) Stratigraphy of the Coranzulí caldera. In: Rocha R, Pais J, Kullberg JC, Finney S (eds) Strati 2013: first international congress on stratigraphy. At the cutting edge of stratigraphy. Springer Geology, Cham, pp 1269–1273
Seggiaro R, Guzmán S, Martí J, Montero C, López E (2014b) La caldera Coranzulí, Puna Norte. XIX Congreso Geológico Argentino, Córdoba, Argentina (Actas CD-Rom Abstracts: S24-2-5)
Seggiaro R, Silvina R, Guzmán S, Martí J (2019) Dynamics of caldera collapse during the Coranzulí eruption (6.6 Ma) (Central Andes, Argentina). J Volcanol Geotherm Res 374:12. https://doi.org/10.1016/j.jvolgeores.2019.02.003

Smith RL (1960) Ash flows GSA. Bulletin 71(6):795–841. https://doi.org/10.1130/0016-7606(1960)71[795:AF]2.0.CO

Soler MM (2005) Caldera Vilama (Mioceno Superior): Su estratigrafía, evolución magmática y relación con eventos ignimbríticos tempranos, Puna Argentina — Altiplano Boliviano. Ph.D. Thesis, Univ. Nacional de Salta, Salta, Argentina

Soler MM, Caffe PJ, Coira BL, Onoe AT, Kay SM (2007) Geology of the Vilama caldera: a new interpretation of a large-scale explosive event in the Central Andean plateau during the UpperMiocene. J Volcanol Geotherm Res 164(1–2):23–53. https://doi.org/10.1016/j.jvolgeores.2007.04

Sparks RSJ, Francis PW, Hamer RD, Pankhurst RJ, O'Callaghan LL, Thorpe RS, Page RS (1985) Ignimbrites of the Cerro Galán Caldera, NW Argentina. J Volcanol Geoth Res 24:205–248

Viramonte JG, Castro Godoy S, Arnosio M, Becchio R, Poodts M (2005) El Campo Geotermal de la Caldera de Cerro Blanco, Utilizacion de Imágenes Aster (resumen), *en* 16° Actas Congreso Geologico Argentino. La Plata: Buenos Aires, Asociacion Geologica Argentina 2:505–512

Viramonte JG, Arnosio M, Becchio R, Gropelli G, Norini G, Corazzatto C, DiFillipo M, Blanco M, Eulillades P, Poodts M, Castro Godoy S, Klotz J, Asch G, Heit B (2005b) Cerro Blanco Volcanic Complex: the youngest caldera system in the Southern Central Andes. A multidisciplinary Earth Science Project, *en* 19° Colloquium on Latin American Geosciences:Potsdam, Alemania, (05/1) 19 LAK, 135

Viramonte JG, Arnosio M, Becchio R, de Silva S, Roberge J (2008) Cerro Blanco Volcanic Complex, Argentina: a late Pleistocene to Holocene rhyolitic arc –related caldera complex in the Central Andes,in: International Association of Volcanologists and Chemistry of the Earth's Interior General assembly, Reykjavik, Iceland: International Association of Volcanologists and Chemistry of the Earth's Interior, 1

Yuan X, Sobolev SV, Kind R, Oncken O, Bock G, Asch G et al (2000) Subduction and collision processes in the central Andes constrained by converted seismic phases. Nature 408:958–961

Chapter 2
Central Composite Volcanoes—Stratovolcanoes

Abstract Cenozoic central composite volcanoes with their steep profiles and conspicuous elevations up to 5 km are distinctive of the Central Andes. The most voluminous and long-lived of those volcanoes are located in the back-arc and extend their activity from 13–10 Ma up to 1.2 Ma. They are andesitic to dacitic in composition, with minor contribution of rhyolites and basaltic andesites. They are the result of Strombolian, Vulcanian eruptions, Pelean episodes involving dome collapse, as well as Plinian and phreatomagmatic events. Distinctive in them are the processes of avalanches of detritus, stages of collapse of craters and collapse of domes, as well as lava stages of construction or reconstruction of the building. Their volcanic products belong to high-K calc-alkaline series with a large compositional range, a broad range of aluminosity, and variable enrichments in incompatible large-ion lithophile elements (LILE) and light rare elements (LREE) and variable degrees of the high field strength elements along their long history. Epithermal deposits of basic metals rich in Ag are frequently related. They correspond to deposits of intermediate sulphuration (IS) with high sulfidation manifestations as siliceous deposits bearing As, Sb, Ag, and in some cases exhalative manifestations of S, in cuspidal portions.

Keywords Cenozoic · Stratovolcanoes · Puna

2.1 Introduction

Cenozoic central composite volcanoes are distinctive of the Central Andes characterized by their steep profiles and conspicuous elevations (they reach up to 5 km). The most voluminous and long-lived composite volcanoes of the Central Andes are located in the back-arc, conforming to the giant calderas, the volcanic landscape of the Puna. They are associated with high-K calc-alkaline magmas, mainly andesitic to dacitic in composition, with minor contribution of rhyolites and basaltic andesites. They are the result of Strombolian, Vulcanian eruptions, Pelean episodes involving dome collapse, as well as Plinians and phreatomagmatic episodes. Distinctive in all of them are the processes of avalanches of detritus, stages of collapse of craters and collapse of domes, as well as lava stages of construction or reconstruction of

B. L. L. Coira and C. E. Cisterna, *Textures, Structures and Processes of Volcanic Successions*, Springer Earth System Sciences,
https://doi.org/10.1007/978-3-030-52010-6_2

the building. They usually precede the caldera formation in their evolution and are also associated with post-collapse stages of the calderas, having in that case smaller dimensions. The development of andesitic, dacitic and, to a lesser extent, andesitic–basaltic or scarce rhyolitic buildings of large to medium size began in the back-arc at 13–10 Ma extending in part of them its activity up to 1.2 Ma.

Their volcanic products belong to high-K calc-alkaline series with large compositional range and show a broad range of aluminosity. Their trace element patterns show variable enrichments in incompatible large-ion lithophile elements (LILE) and light rare elements (LREE) and variable degrees of the high field strength (HFS) observed along their long history (Kay et al. 2010; Richards et al. 2006).

Epithermal deposits with vetiform and stockwork of basic metals rich in Ag are frequently related to the stratovolcanoes of the region. They correspond, to a large extent, to deposits of intermediate sulphuration (IS). In some cases, high sulfidation levels are preserved in the cuspidal portions with partially breached siliceous deposits and anomalous contents in As, Sb, Ag (geothermal discharge zones). At these levels, exhalative manifestations of S can be identified. Deposits carrying Sb-Mn mineralization, and with anomalies in Ag, Au, Bi, Mo, W are marginally arranged and represent paleoaquifers and distal thermal sources.

In this chapter will be analyzed the morpho-structural, compositional, and evolutionary characteristics of these distinctive volcanic buildings through selected examples.

Compositional variations, magmatic processes, distribution, and dimensions of the Southern Central Andes Stratovolcanoes are excellent records for understanding the geodynamic evolution of the region.

2.2 Quevar Volcanic Complex

The Quevar Volcanic Complex (24°20′S: 66°45W, 6180 m.s.n.m.) is one of the biggest composite volcanoes of the Puna (550 km^3, Goddart et al. 1999), built on ignimbrites from the second Cerro Aguas Calientes caldera cycle (10.3 Ma; Petrinovic et al. 2010). The center has been dated and studied by Goddard et al. (1999). It was emplaced in the lapse 10–5 Ma (Olson and Gilzean 1987) along major NW–SE trending wrench corridor (Calama-Olacapato-El Toro lineament, see Fig. 2.1) that controlled the location of the conspicuous and large transversal volcanic chain, of which it is part, and extends to Chile. High-angle left-lateral strike-slip faults parallel to that lineament are dominant throughout the edifice and act to pull apart the intervening strata forming NE-SW normal faults and have controlled the alignment of the main emission mouths.

The Quevar stratovolcano contains several eruptive centers which are mainly composed of andesitic to a lesser extent dacitic lava flows, small rhyolitic domes containing perlithic deposits and block and ash deposits.

Fig. 2.1 Satellite images of Quevar Volcanic Complex

The activity of the Quevar complex began with a rhyolite lava dome (8.6 Ma), brecciated and perlitized, erupted on the south flank, to the west of the Cerro Azufre. This dome was then covered in large part by major andesitic–dacitic lava flows and minor block and ash deposits that erupted from five vents (see Fig. 2.1): Cerros Quevar (Fig. 2.2a), Gordo, Azufre (Fig. 2.2b), and Mamaturi and Quironcolo, during the lapse 8.3 to 7.53 Ma (Goddard et al. 1999). The initial eruption of this group took place from a vent to north of Cerro Gordo. It was represented by a dacite with large sanidine and quartz phenocrysts (Fig. 2.2c) and with frequent mafic inclusions (hornblende, plagioclase, bordering olivine, and augite).

Activity then spread to the SE to form the andesitic cones of Cerros Mamaturi and Quironcolo corresponding to andesites with hornblende and biotite (Fig. 2.2d)

Fig. 2.2 **a** Quevar Volcanic Complex, on the right Cerro Quevar. **b** Cerro Azufre on the left and Cerro Gordo on the right. **c** Cerro Quevar Dacite Dome. Phenocryst of sanidine (up to 6 cm in diameter), quartz, biotite, and hornblende. **d** Andesite lava flow from Cerro Azufre. Phenocryst of plagioclase, hornblende, and biotite. Plane polarized light

and without mafic inclusions unlike N lavas. The largest event from the Quevar was erupted from a vent to the north of Cerro Azufrero (Azufrero Andesite) and covered a substantial portion of the Complex. Then the activity returned to Cerro Quevar with the emission of the Esquina Azul Andesite that flowed to the N and NE and finally with the formation of a dacitic dome (5.5 Ma) that crops out in the peaks Gordo and Quevar, characterized by the great development of sanidine phenocrysts up to 5 cm in diameter (Goddard et al. 1999). Important hydrothermal alteration and epithermal deposits of Ag, Pb, Zn (Sb, As, Bi) (Olson and Gilzean 1987; Goddard et al. 1999; Petrinovic et al. 1999; Robl et al. 2009) occur along northeast- and west-northwest-trending faults in the complex (see Fig. 2.1 Quevar North and South alteration areas). Those alteration zones are located on the flanks of the volcano, where physical erosion by glaciation and sector collapse expose deep parts of the sub-volcanic system. An example of the important deposit identified within the Quevar South alteration zone is the Yaxtché deposit (Fig. 2.3a, b) as part of El Quevar Project (Golden Minerals Co. 2018). This deposit is located in an epiclastic unit consisting of a matrix-supported volcanic breccia intruded by a complex of porphyritic dacite domes and associated breccias and flows. Silver is the element of economic significance in the deposit, and anomalous concentrations of copper, lead, zinc, and lesser gold occur locally.

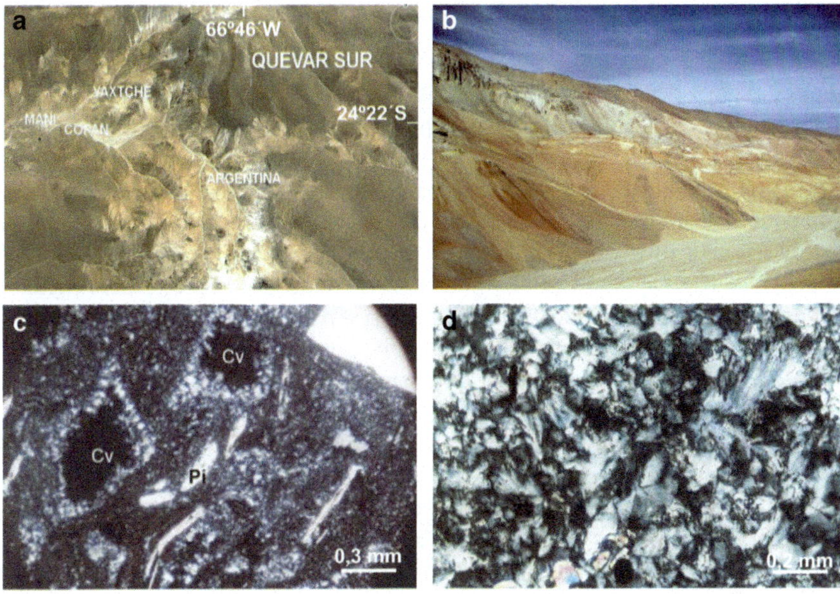

Fig. 2.3 **a** Quevar Sur Project. Main mineral deposit, after Golden Minerals Co. **b** Yaxtche deposit. **c** Vuggy silica (Cv) after feldspar. Matrix replaced by dickite, kaolinite, and quartz in hydrothermal alteration zone Crossed polarized light. **d** Alunite–dickite aggregate. Steam-heated sector in the uppermost portions of Cerro Quevar. Crossed polarized light

The sulfide (hypogene) mineralization is represented by pyrite, galena, sphalerite, tetrahedrite, tennantite, complex Pb Sb Bi $\pm$ Ag sulfosalts, bismuthinite, stibnite, and chalcopyrite; oxides (supergene): plumbojarosite, argentojarosite, limonite, and stibiconite. Zoned advanced argillic alteration is typical of that which might be expected to occur in association with high-sulfidation epithermal gold deposits (Golden Minerals Company 2018). Elements of the advanced argillic alteration zone from the center outwards: vuggy silica (Fig. 2.3c), pervasive silica, silica–alunite, kaolinite–dickite, neutral argillic, steam-heated: (alunite–dickite Fig. 2.3d), propylitic. The Yaxtché deposit alteration assemblages are typical of high sulfidation epithermal deposits, whereas the metal content and sulfide assemblages are characteristic of mineralizing fluids with an intermediate sulfidation state (Golden Minerals Company 2018).

The ore mineralogy may correspond to the most southern part of the Ag-Sn metallogenetic province of Bolivia (Robl et al. 2009).

2.3 Antofalla Volcanic Complex

The Antofalla Volcanic Complex (AVC) is a long-lived central composite volcano of 35 km in diameter and 6400 m maximum height located on the Southern Puna plateau of NW Argentina (25°35′S–67°55W), ~100 km east of the volcanic arc (Figs. 2.4 and 2.5). It lies on the NW–SE Archibarca lineament, one of several transverse structures that cut across the Andes from Chile. The AVC has a protracted magmatic history from 13 to 1.6 Ma (Coira et al. 1993; Kay et al. 2008). Initial volcanism was characterized by eruption of voluminous rhyolitic ignimbrites. These rhyolites record significant geochemical evidence for crustal contamination, or derivation by crustal melting, which decreases upwards through the pyroclastic sequence. They were followed by extrusion of potassic lavas of the shoshonite–latite–trachydacite suite from 10.1 to 9.5 Ma. Subsequent magmatism (9.1–1.6 Ma) was of basaltic andesite–andesite–dacite composition, with localized, smaller volume, dacitic and rhyolitic ignimbrites.

Monogenetic basaltic andesite cinder cones and lava flows were erupted in the Quaternary in response to a change to transtensional tectonics in the Puna.

Its early volcanic records (Early Phase) correspond to lavic central buildings constituted by andesites–pyroxene basaltic andesites–(opx, cpx) and minor with pyroxene–olivine.

They correspond to the Cerros La Aguada (12.9 ± 0.8 Ma, Kraemer et al. 1999), Cajero (13.0 ± 0.5 Ma Schnurr 2001), and the Antofalla small cone (13 Ma, Coira and Pezzutti 1976). They are succeeded by olivine basaltic flows with opx or

Fig. 2.4 Satellite images of the Antofalla Volcanic Complex

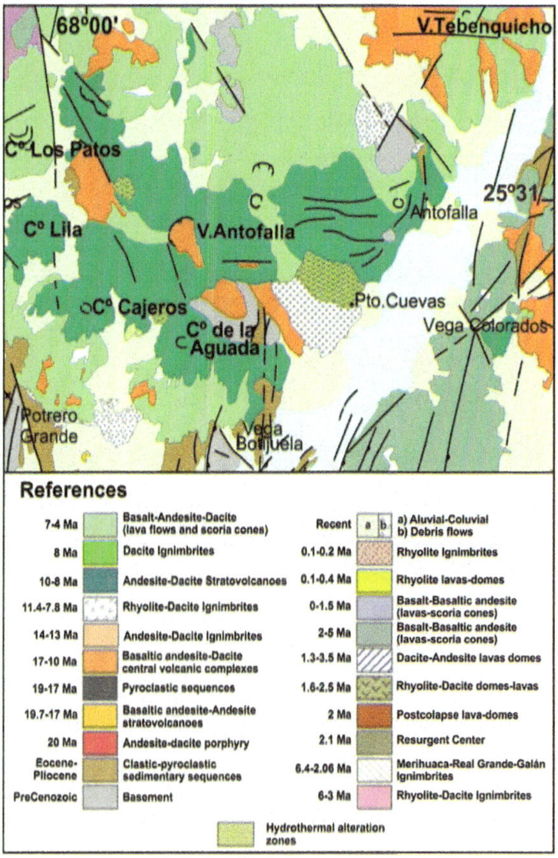

Fig. 2.5 Geological map and satellite images of the Antofalla Volcanic Complex

cpx poorly exposed (10.5 Ma) as well as frequently dislocated and partially hydrothermalized glassy rhyolites (11 Ma). During the lapse 10.9–9.6 Ma an explosive event occurred, represented by dacitic–rhyolitic ignimbritic deposits with biotite (Fig. 2.6b) of valley-fill type, which are recognized in the marginal or strongly dissected area of the volcanic complex, as in the Qda. Antofalla where 3 members are separated by short clastic episodes (Antofalla Ign 1, 2, 3, Schnurr 2001).

Thus they also extend to the N, covering the granite of Antofalla and extending to the north of the Volcano and south of the Salar de Archibarca, above the Tertiary sedimentary deposits. Following and partly in contemporary form (10–9 Ma,) volcanic-domic centers, such as those of the Cº Lila (Coira and Pezzutti 1976), were built. They were widely disseminated, forming a large part of the substrate of the Antofalla Volcanic Complex and partially drowning the set of central buildings of the Early Phase (Fig. 2.6a). They correspond to volcanic centers constituted by dacitic lavas with biotite and minor opx and cpx and frequent xenocrystals of Q or FK +Pl (Fig. 2.6c). These emissions were followed by important and repeated andesitic lava effusions (with hbl, cpx, and opx) and hornblendiferous dacitic flows with opx quartz,

Fig. 2.6 **a**. Antofalla Volcanic Complex at the foot of the Salar de Antofalla. **b** Dacitic ignimbrite with plagioclase, quartz, and biotite phenocrysts, moderately welded. Plane polarized light. **c** Dacite with hornblende opx quartz and biotite. Crossed polarized light. **d** Hornblendiferous dacite with quartz, biotite < ortopyroxene. Crossed polarized light

biotite (Fig. 2.6d) with lower participation toward the end of this event (9–8 Ma) of andesites to pyroxene basaltic andesites with hbl (Fig. 2.7a), those that make up the main structure of the Antofalla volcano (Coira and Pezzutti 1976). The following lavic episode (7–4 Ma) was recorded in marginal areas of the Volcanic Complex, as flank vents with strong structural control. It was represented by hornblendiferous pyroxene basaltic andesites (opx + cpx) (Fig. 2.7b) and by olivine, augite + opx basaltic andesites and minor olivine basalts with augite (Fig. 2.7c). It conformed cones of scoria related to monogenic eruptions, as well as fields of lava as a result of the superposition of different flows. In 6.6–4.3 Ma took place at the E and S of the Antofalla Salar important emissions of basaltic andesites, basalts with olivine and pyroxene and pyroxene andesites (opx + cpx), together with occasional dacites.

At 2.5–2.3 Ma took place the eruption of reduced degassed glassy domic rhyolitic lavas and domes controlled by fracturing W-E (Schnurr 2001; Richards et al. 2006). Monogenic andesitic–basaltic scoria cones and lava flows, associated with extensional faulting systems, common to the E and S of the Salar de Antofalla, were almost absent in the Antofalla volcano, only scarce representatives being recognized in the last 1.5- 1.2 Ma (Marrett et al. 1994; Kay et al. 1997), when the eruptive activity ended in the Antofalla volcanic complex.

Epithermal deposits of Au and Pb–Zn–Ag occur within the AVC, apparently related to the formation of rhyolite domes (Quebrada de las Minas, Fig. 2.7d). The

Fig. 2.7 **a** Pyroxene basaltic andesites with hornblende. Crossed polarized light. **b** Pyroxene basaltic andesite hornblendiferous. Crossed polarized light. **c** Olivine basalts with augite. Crossed polarized light. **d** Quebrada de Las Minas. Rhyolite dome. Hydrothermal alteration area

mineralization style is typical of other Andean volcanic centers, and may be underlain by deeper seated porphyry styles of mineralization (Richards et al. 2006). Extensive blanketing by younger lavas and limited erosion restrict exposure of these altered and mineralized sequences, however, although the potential for further discoveries beneath cover or at depth in this system, appears high.

2.4 Rachaite Stratovolcano

Rachaite stratovolcano (8.4–7.2 Ma) is located (22°53′S, 66°08′W) northeast of the Coranzuli caldera to which it precedes (see Fig. 2.8) and at the eastern margin of the Northern Puna. This center has been studied by Coira (1979a, b), Coira et al. (2004) and Heidorn (2002).

It is a high-K calc-alkaline central composite volcano structurally controlled by the NW–SE Lipez-Coranzuli lineament, a transverse structure that cut the Andes from Chile. (Fig. 2.9).

Its **Initial stage** (8.4 Ma, Heidorn 2002) was represented by the eruption of andesitic to dacitic ignimbrites (Fig. 2.10a), ash-fall deposits (Fig. 2.10b), domes, and associated block and ash deposits (Fig. 2.10c). They were followed by reworked

Fig. 2.8 Satellite image of Rachaite stratovolcano

pyroclastic-rich deposits (Fig. 2.10d) and laharic events. The **second stage** (8.2–7.9 Ma, Heidorn 2002) consisted in the cone building, resulting from the effusion of andesitic (Fig. 2.10e, f) and minor basaltic andesitic lavas.

The **third stage** consisted in the partial collapse of the volcanic building with the formation of a crater structure preserved in the eastern sector, meanwhile the edifice is partially eroded in its nucleus and drowned by the ignimbritic flows (6.6 Ma, Seggiaro 1994) associated with the Coranzuli caldera (see Fig. 2.8).

The **fourth stage** was represented by the emplacement of andesitic dome and dykes (Figs. 2.10f and 2.11b) (7.8–7.2 Ma, Heidorn 2002) exposing a mineralized stock after collapse sector.

An active hydrothermal system created a high-sulphidation system in the uppermost sections of the mineralized structure, partially eroded, and at lower levels an intermediate-sulphidation system: polymetallic Chocaya deposit (Fig. 2.12). Low-sulphidation deposits formed in distal sectors from fluids that discharged as hot springs on the outer slopes of the Rachaite stratovolcano (e.g., Doncellas–San José:

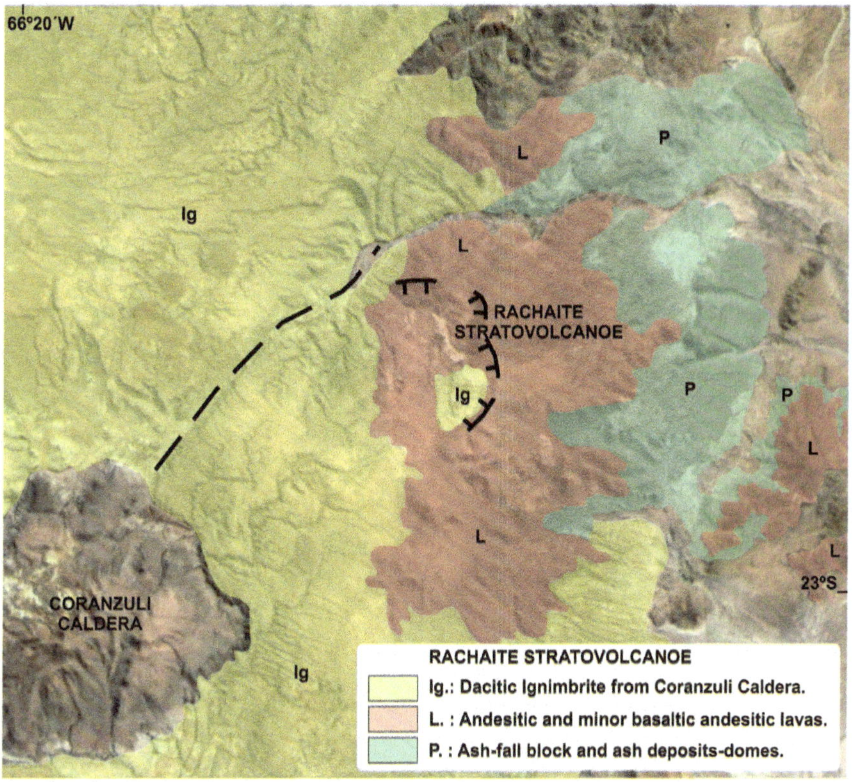

Fig. 2.9 Rachaite stratovolcanoe stratigraphy and mineral deposits associated

Sb-Fe; Pabellón–Iral: Fe–Mn; Coira and Brodtkorb 1995) (Fig. 2.13). The Pb–Zn-Ag mineralization of the Chocaya deposit is centered within the collapse crater and exposes ~6 km^2 of phyllic and propylitic alteration, silicification, and younger steam-heated advanced argillic alteration (Fig. 2.14) (Coira et al. 1990). Ore minerals (galena, sphalerite, tetrahedrite) metal base/precious 100:1; Ag/Au > 1, gangue (quartz, carbonate, barite), mineralization textures, and type of alteration lead to characteristics of intermediate-sulphidation state (Heidorn 2002) (Fig. 2.15).

2.5 Tuzgle Volcano

Tuzgle volcano (24°S, 66° 30′W) with an altitude of 5486 m a.s.l. sits near the eastern edge of the Puna plateau, about 275 km behind the main volcanic front of the CVZ and ~200 km above the Wadati–Benioff zone (Fig. 2.16).

Fig. 2.10 **a** Dacitic ignimbrite rich in pumice. **b** Air-fall tuff rich in pumice. **c** Block and ash deposit unsorted with cognate andesitic blocks which reach up to 30 cm in diameter, some of them with cooling joints indicative of their hot emplacement. **d** Reworked pyroclastic-rich deposits with andesitic lithics. **e** Cuspidal andesite lava flows (8.2 Ma). **f** Andesite dyke post hydrothermal alteration (7.2 Ma)

The volcano is located in a N-S-trending, thrust fault-bounded, intramontane depression (Fig. 2.17). Volcanic activity at Tuzgle back-arc volcano postdated extensive Miocene to early Pliocene andesitic and dacitic eruptions from large stratovolcanoes like Queva, south of Tuzgle (Fig. 2.16). Volcanic activity began at 0.5 Ma (Aquater 1979) in this central composite volcano with the eruption of ~0.5 km^3 of rhyodacitic ignimbrite which flowed to the north over Miocene volcanic rocks and the Ordovician Basement and formed a plateau of 60 km^2, 2–80 m thick.

Subsequently, at 0.3 Ma (Aquater 1979) a complex of dacitic lava dome were erupted (Old Complex) with a total volume of about 3.5 km^3. Part of this dome complex was then covered by andesitic flows of the pre-Platform Unit, which flowed

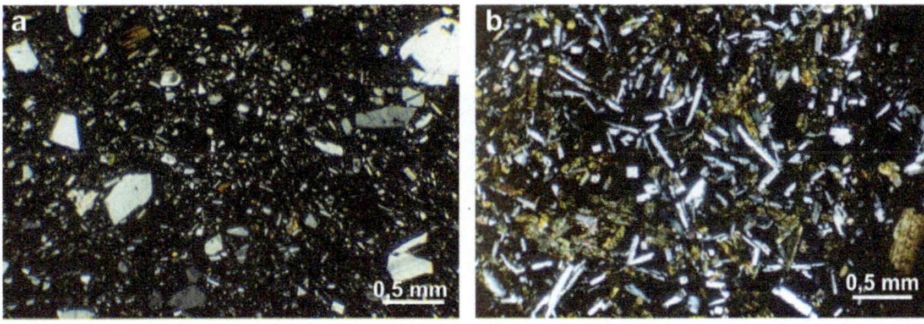

Fig. 2.11 **a** Andesite with phenocrysts of plagioclase, micro-phenocryst of pyroxene, and hornblende in a hialopilitic groundmass. Crossed polarized light. **b** Andesite with phenocrysts of lamprobolita and biotite. Crossed polarized light

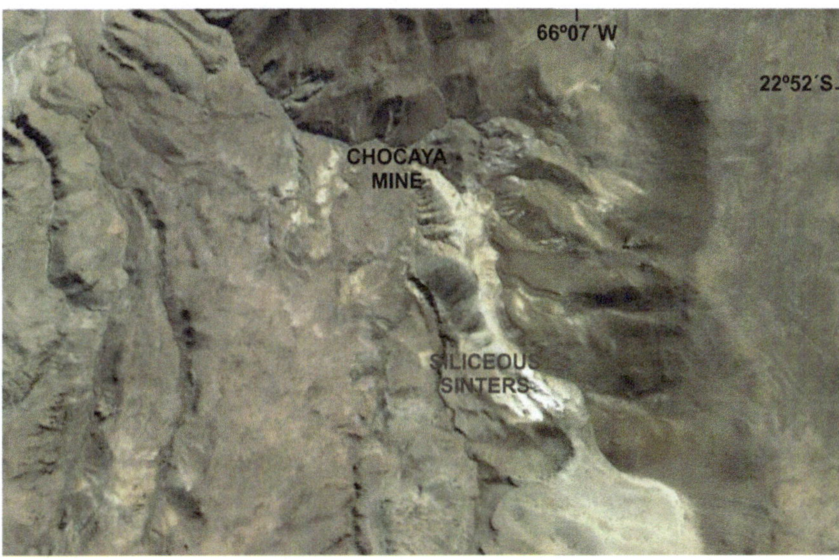

Fig. 2.12 Chocaya Mine (Pb, Zn, Ag) and cuspidal siliceous sinter (hot-spring discharge)

in NNE-SSW direction from a central source area, partially delimited by a volcano tectonic scarp (crater rim) (Fig. 2.18). Associated with that scarp, Norini et al. 2014 estimated in ≈ 0.5 km^3, the volume of the correspondent sector collapse of the edifice. The next event corresponded to the emission of mafic andesite lavas (Platform Andesite Unit) which flowed covering a wide flat surface (unconformity) at the top of the volcano.

Then tectonic reactivation occurred in the sector with east–west and northwest-southeast faulting, the one which controlled the eruption of the lava flows of the post-Platform and the Recent Young Flow Unit (Coira and Paris 1981). The total

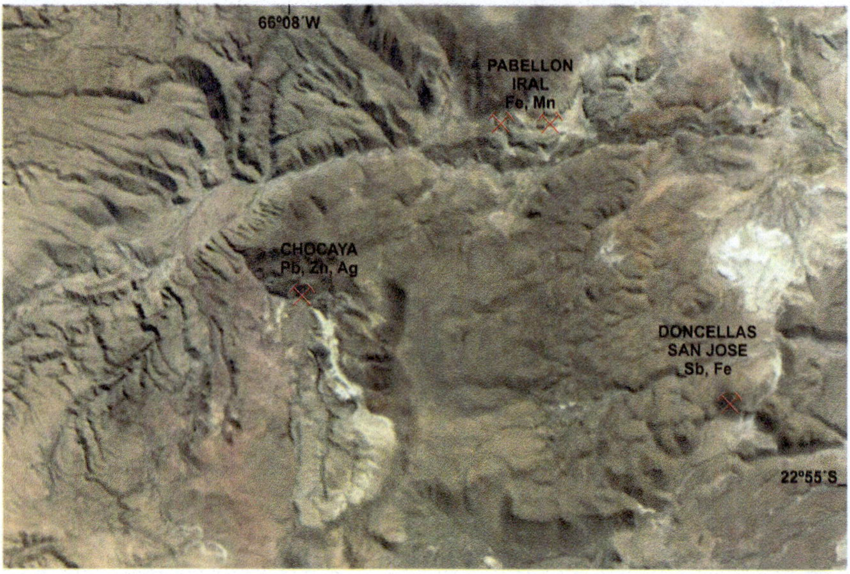

Fig. 2.13 Polymetallic Chocaya deposit and low-sulphidation deposits from hot springs on the outer slopes of the Rachaite stratovolcano

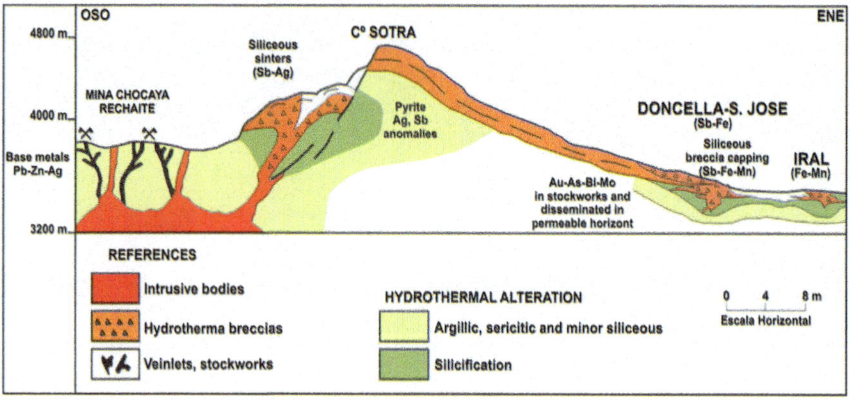

Fig. 2.14 Sketch of mineral deposits distribution on Rachaite stratovolcano (Coira and Brodtkorb 1995)

volume emitted in the effusive central stage, post-Old Complex, has been estimated at 0.5 km^3 (Aquater 1981).

The Tuzgle Ignimbrite substrate of the lava edifice (Fig. 2.19a) is dacitic in composition, relatively crystal-rich (20–25%), and shows small to moderate welding (Fig. 2.19b).

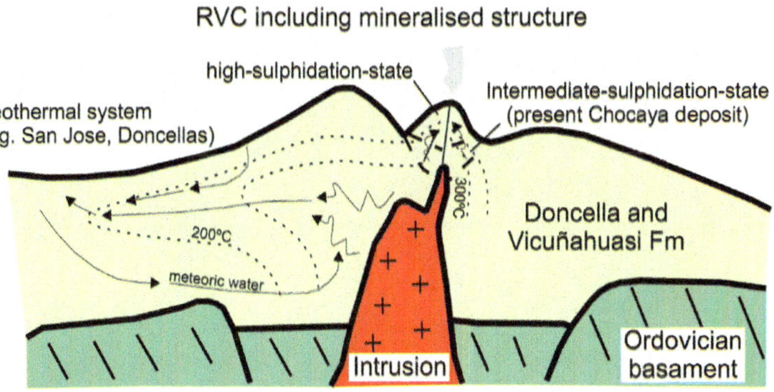

Fig. 2.15 Model of hydrothermal system of Rachaite stratovolcano after Heidorn (2003)

Fig. 2.16 Satellite images of Tuzgle region

Fig. 2.17 Satellite images Tuzgle volcano

Lithic fragments are found at the base, while pumice fragments constitute 10–20% of the middle and upper levels. In order of abundance, phenocrysts are normally zoned plagioclase (An45-32), quartz, biotite, and alkali feldspar,,titanomagnetite. Compositional similarities between these phenocrysts and xenocrysts in the Tuzgle lavas suggest that the xenocrysts were derived from a host like the ignimbrite.

The lavas that make up the composite central Tuzgle volcano (Figs. 2.20 and 2.21) are crystal-rich andesite and dacite with blocky structures. Their most striking feature is the contrast between their large, feldspar, quartz xenocrysts and their biotite grains, and small mafic phenocrysts and hyalopilitic to pilotaxitic groundmass. The groundmass phases include plagioclase, olivine, clinopyroxene, orthopyroxene, opaque oxides, and apatite. Others notable characteristics are (a) absence of plagioclase phenocrysts, and (b) generally regular variation of mafic phenocryst types (olivine, phlogopite, clinopyroxene, orthopyroxene, and amphibole) with whole rock composition. In all these rock types, the most common xenocrysts are plagioclase (An_{33} to An_{38}) and quartz. Disequilibrium with the host magma is indicated by sieve textured, calcic rims (to An_{76}) on plagioclase (Fig. 2.22a), and embayed brown glass rims and clinopyroxene (Fs_{15} En_{45}–Wo_{40}) on quartz (Fig. 2.22c). These features are best developed in the most mafic rocks.

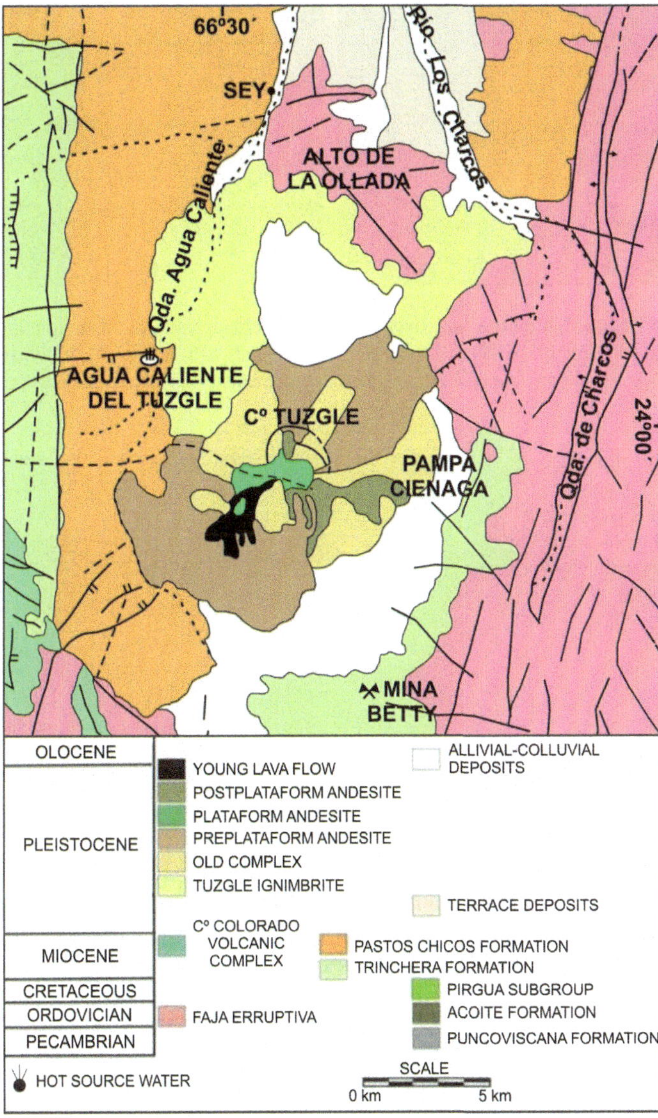

Fig. 2.18 Tuzgle volcano and its stratigraphy, modified from (Coira 1995)

A few fragmental alkali feldspar grains (Or_{56} to Or_{76}) occur in andesites pre-Platform and post-Platform and in Old Complex dacitic rocks.

Mafic phases in rocks with >59% SiO2 are variable. Enstatite, augitic clinopyroxene, and minor amphibole occur as microphenocryst and in the groundmass in Andesite: pre-Platform and post-Platform (Fig. 2.22b, c, d). Pargasitic amphibole micro-phenocrysts occur in andesite and dacite (Old Complex Unit). Meanwhile

Fig. 2.19 **a** Tuzgle Ignimbrite, massive moderately welded with columnar jointing. **b** Tuzgle Ignimbrite, dacitic in composition, crystal-rich and welded, with alignment of elongated pumice and phenocrystals (plagioclase (An $_{45-33}$), quartz, biotite, and alkali feldspar). Plane polarized light

Fig. 2.20 Tuzgle edifice. View of Old Complex, cuspidal post-Platform and Platform lavas, succeeded by the young andesitic flow of Tuzgle volcano

mafic phenocrysts in andesite with <59% SiO2 (Platform and Young Flow Units) include euhedral to subhedral grains of salitic clinopyroxene and forsteritic olivine (Fig. 2.23a, b) . Coira and Kay (1993) reported thermobarometric data suggesting a >2.5–3 kbar pre-eruption crystallization pressure for the oldest unit of the Cerro Tuzgle volcanic succession, the Tuzgle Ignimbrite, and a 5–6 kbar pre-eruption crystallization pressure for the overlying lava units.

Tuzgle volcanic rocks follow a high-K andesite to rhyodacite trend on a KzO versus SiO2 diagram and have calc-alkaline arc-like A12O3 concentrations (15–17%) and FeO/MgO ratios (1–2.5). Trace element data, particularly REE data, can be used to separate the Tuzgle suite into an older and a younger group; the older, low La/Yb group consists of the Tuzgle Ignimbrite, the Old Complex, and the pre-Platform, Platform, and some of the post-Platform units. These samples have moderate to steep REE patterns (La/Yb ratios = 20 to 28) that tend to steepen as SiO_2 concentrations increase from 56 to 70%. The younger, high La/Yb group consists of the other part of the post-Platform unit and the Young Flow. These samples have 58–61% SiO2 and steep REE patterns (La/Yb ratios > 35). The most mafic sample of this group is the Young Flow with 59.9% SiO_2 content. Differences between patterns in mafic

Fig. 2.21 Detail of the Young lava flow showing blocky structure

lavas can be attributed to variable REE retention by garnet during separate melting events producing discrete precursor magmas (Coira and Kay 1993). Another feature of Tuzgle REE patterns is that negative Eu anomalies (Eu/Eu* = 0.78 to 0.56) occur in all samples, whether or not they contain feldspar phenocrysts. These anomalies increase in size with increasing whole rock SiO2 concentration and are attributed to mixing of feldspar-free mafic magmas with silicic magmas that have fractionated feldspar (Coira and Kay 1993). A silicic magma related to the Tuzgle Ignimbrite, in which the xenocrysts would have been phenocrysts, is an obvious choice.

Petrologic and geochemical data show that volcanic rocks from Tuzgle volcano are complex mixtures of crustal and mantle-derived components. The Tuzgle rocks can be divided into two eruptive cycles with different mantle precursors: (a) an older, more voluminous dacitic to mafic andesitic sequence with high La/Yb ratios; and (b) a younger andesitic sequence with higher La/Yb ratios. Successively more silicic units contain greater amounts of silicic crustal material. Some young lavas contain a shoshonitic-like component (Coira and Kay 1993). Considering the petrologic and geochemical characteristics of Tuzgle volcanic rocks, (Coira and Kay 1993) proposed a model (Fig. 2.24) in which it is contemplated as (1) melting in the asthenospheric mantle above the subducted plate, (2) melt contamination by a delaminated lithosphere modified by plate in the asthenosphere and by a continental crust and lithosphere in situ, (3) accumulation of magma, fractionation, and mixing of various mafic magmas with the dacite-rhyolite that produced the ignimbrite, at the depth

Fig. 2.22 **a** Andesite showing plagioclase xenocrysts sieve textured, with calcic rims (to An_{76}) Crossed polarized light. **b** Andesite pre-Platform with micro-phenocrysts of amphibole and clinopyroxene in a hyalopilitic to groundmass. Crossed polarized light. **c** Andesite pre-Platform with micro-phenocrystals of enstatite and augite and xenocrystals of quartz embayed by brown glass and clinopyroxene rims Crossed polarized light. **d** Andesite post-Platform with phenocryst of augitic clinopyroxene, plagioclase in a pilotaxitic groundmass. Crossed polarized light

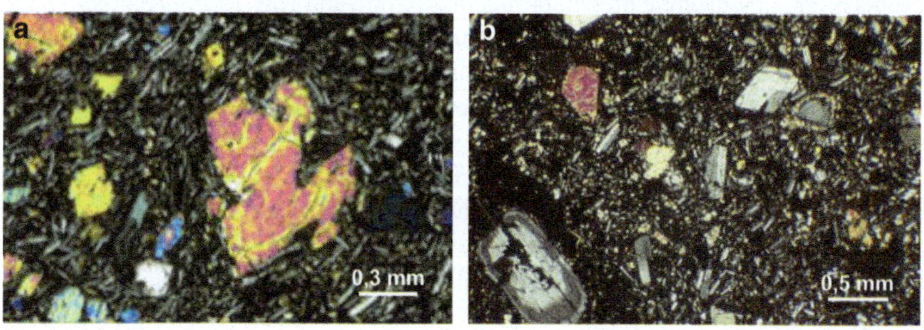

Fig. 2.23 **a** Platform Andesite (<59% siO2) with forsterite phenocryst and micro-phenocryst of clinopyroxene immersed in the glassy groundmass which includes laths of plagioclase. Crossed polarized light. **b** Andesite with olivine, clinopyroxene, ortopyroxene phenocrystals, xenocrystals of sieved plagioclase, embayed quartz by glass and clinopyroxene in a hyalopilitic groundmass and opaque oxides in a scarce glass matrix Crossed polarized light

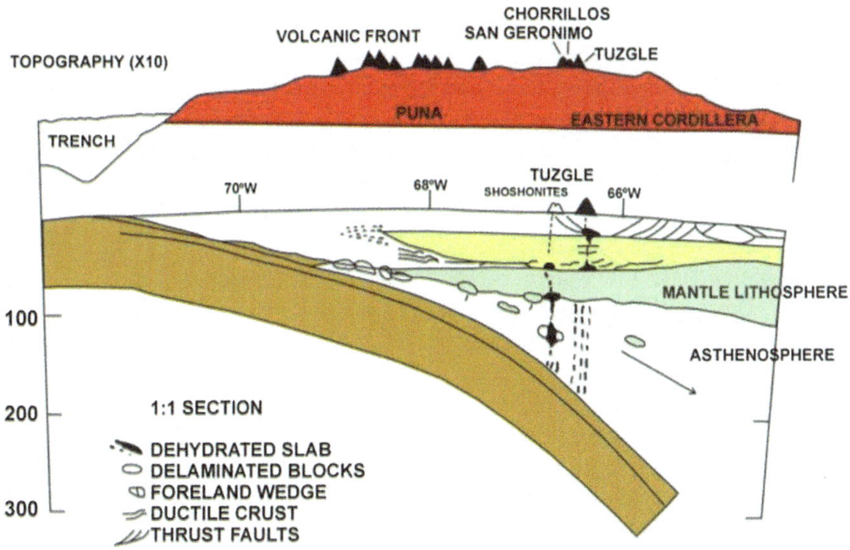

Fig. 2.24 Sketch of tectomagmatic model for Tuzgle volcano (modified from Coira and Kay (1993))

of the decollement (about 20 km) before the eruption. Schurr et al. (2006) on the basis of a seismic tomographic model interpreted the presence below the Tuzgle of the continental lithosphere block in the process of delamination, fusion in the lower crust, and melt accumulation at ~20 km, in accordance with the model proposed by Coira and Kay (1993).

A geothermal field is associated with this central composite volcano in which hot springs are recognized with temperatures of 40° to 56 °C in the margins of the Aguas Calientes river (Figs. 2.18 and 2.25a) and 20 °C in the Betty Mine Plant. Deposits of siliceous sinters (opalo-chalcedony) laminar and brecciated, as well as isolated aggregates of boronatro-calcite associated with the thermal area of Tuzgle Aguas Calientes have been recognized (Fig. 2.25b).

Fig. 2.25 **a** Tuzgle thermal baths. **b** Deposits of siliceous sinters and isolated aggregates of boronatro-calcite related to the thermal area of Tuzgle Aguas Calientes

Geothermal fluids up well in areas with more intense fracturing, especially where main regional structures, particularly NW–SE, parallel to Calama–Olacapato–Toro lineament, intersect with N-S faults and secondary structures such as the extensional ENE-WSW faults (Giordano et al. 2013).

Through the different phases of the studies carried out (Aquater 1981; CREGEN 1988; Coira 1995), it was determined that the waters in the Aguas Calientes Quebrada are alkaline chlorides, with high specific conductivity (2300 a 6600 umho/cm), intermediate values in the Ammonium content (4.3–12.6 mg/l), and high SiO2 content (58.2–148 mg/l). Its low water/gas ratio indicates the presence of a "hot water type system". These geochemical studies indicated mixtures between deep and superficial water and temperatures for the reservoir from 132 °C to 142 °C, according to the geothermometers K Mg and chalcedony, respectively.

In the Tuzgle volcano area, two high-gradient zones, one located to the south and one to the northwest of the volcano, could be established as a result of a thermal gradient well-drilling program (Coira 1990). On the basis of geophysical studies carried out (vertical electric soundings, Hidroproyectos -SETER-CEPIC, 1984, CREGEN 1988) was recognized in the area of the Tuzgle Volcano a possible shallow reservoir at 80–150 m depth (Trincheras Formation) with a thickness of 50 to 300 m and another possible reservoir, below 340–670 m, with a thickness of 350–650 m which could correspond to Ordovician magmatic-sedimentary sequences and which would correspond to the High Conductive Electro-stratum determined by the Geophysical Area Engineering S.A, 2015, in its geoelectrical prospecting of the area (see Fig. 2.26). They also detected a High Resistive Electro-stratum at 870–1200 m which could correspond to the Precambrian–Cambrian Basamento. On the other hand, a third possible reservoir has been suggested by the Geophysical Area

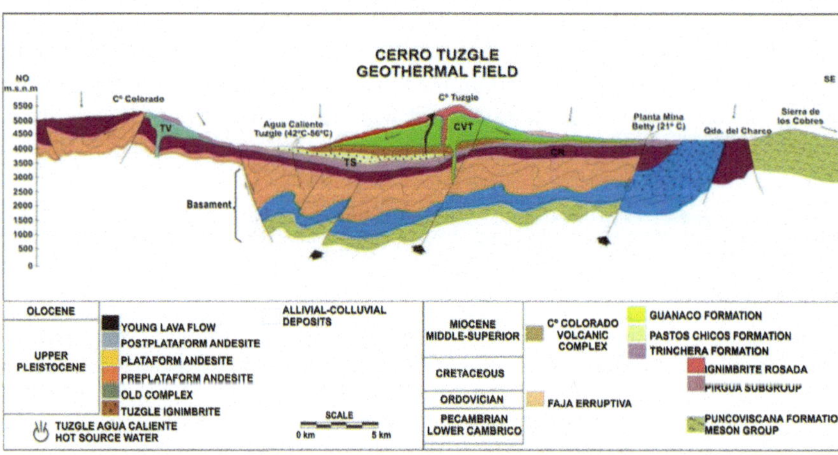

Fig. 2.26 Circulation scheme of geothermal fluids in the Cerro Tuzgle area. Modified from Coira (2008)

Engineering S.A as a second High Conductive Electro-Formation, below the High-Resistive Formation. Likewise Sainato and Pomposiello (1997), in their magneto-telluric 2D model of distribution of resistivity, identified two main bodies. The first one represented by a conductive layer, of low density, with its top at 2 km and its base at ~3.1 km of depth, coinciding with the depth reported for the Basement. The second identified body of 0.5 Ω-m of resistivity was recognized surrounded by a more resistive zone (100 Ω-m), with a density contrast = 0.04 g/cm^3.

The latter is related to a possible magmatic chamber surrounded by saline fluids, with its roof at ~8 km and its base at ~22 km depth. Characteristics these last ones consist with the proposal on petrological bases were realized by Coira and Kay (1993) for the region of accumulation of magmas.

In the scheme of circulation of geothermal fluids to the west of Cerro Tuzgle (Fig. 2.26), hot and cold waters are separated by an impermeable layer (Pastos Chicos Formation), while at E there is a mixture between deep and surface waters, when missing partially that layer. The recharge areas resulting from the infiltration of snowfields are located in the southern, eastern, and western areas, controlled by regional fracturing. The recharge altitude was determined at 4,400–4,500 m s.n.m. The high topography, parallel to the Toro line, interrupts the geothermal field of the Tuzgle to the south, as denoted by the geochemical differences between the Tuzgle and Tocomar geothermal systems. The major and most frequent lineaments in the region have a NW-SE, ENE-WSW and N-S strike and correspond to the regional tectonic structures that have controlled both regional recharge and the rise of thermal fluids.

References

Aquater (1979) Estudio del potencial geotérmico de la provincia de Jujuy, República Argentina, Fase de reconocimiento.Secretaría de Estado de Minería: 129 pág. Informe inédito. Buenos Aires

Aquater (1981) Exploracion geotermica del/trea del Cerro Tuzgle.Provincia de Jujuy, República Argentina. Secretaria de Estadode Mineria, Argentina, open-file report

Coira B (1979a) Descripción Geológica de la Hoja 3C Abra Pampa, Provincia de Jujuy. Servicio Geológico Nacional Bol 170:1–90

Coira B (1979b) Episodios volcánicos cenozoicos en el área de Chajarahuaico-Casabindo, Puna jujeña. Actas VII Congreso Geológico Argentino II:537–551

Coira B (1990) Informe preliminar sobre el programa de perforaciones a fin de determinar gradientes de temperatura en el área del Cerro Tuzgle. Dirección Provincial de Minería y Recursos Energéticos de Jujuy- Universidad Nacional de Jujuy- Dirección de Energía de Jujuy, pp 55 unpublish report

Coira B (1995) Cerro Tuzgle Geothermal Prospect, Jujuy, Argentina. Proc World Geotherm Congr 2:1161–1165

Coira B.y Brodtkorb M (1995). Polymetallic mineralization associated with Cenozoic volcanism in Northern Puna Argentina. In: Mank JL, George JDS (eds) Proceeding of The PACRIM congress, pp 135–140. Auckland-New Zealand

Coira B, Malhburg Kay S (1993) Implications of Quaternary volcanism at Cerro Tuzgle for crustal and mantle evolution of the Puna plateau, Central Andes, Argentina. Contrib Mineral Petrol 113:40–58

Coira B, Paris G (1981) Estratigrafia volcfinica del area del Cerro Tuzgle (23 ° 50' S y 66 ° 45' W), Provincias del Jujuy-Salta. VIII Cong Geol Argen III:659–671

Coira B, Pezzutti N (1976). Vulcanismo cenozoico en el ámbito de Puna catamarqueña (25°30'-25°50' Lat.S y 68°-68°30' Long.O). Asociación Geológica Argentina Revista XXXI (1):33–52

Coira B, Chayle W, Barber E, Solís N, Brodtkorb M, Camacho M, Díaz A (1990) Paleosistema geotermal del Terciario superior y su mineralización de metales básicos (Pb, Zn, Ag), Rachaite, Jujuy, Argentina. Actas XI Congreso Geológico Argentino, I: 303–306

Coira B, Malhburg Kay S, Viramonte J (1993) Upper cenozoic magmatism of the Argentina Puna. Int Geol Rev 35(8):677–720

Coira B, Caffe P, Ramírez A, Chayle W, Díaz A, Rosas S, Pérez A, Pérez B, Orozco O, Martínez M (2004) Descripción de la Hoja Geológica 2366-1 "Mina Pirquitas". Boletín del Servicio Geológico Minero Argentino Nº 269:1–123

Coira B (2008). Recursos geotérmicos de alta entalpía de la Provincia de Jujuy. En Coira B, Zappettini EO (eds) Geología y Recursos Naturales de la Provincia de Jujuy, Relatorio del XVII Congreso Geológico Argentino. Asociación Geológica Argentina. Buenos Aires, pp 563–576

CREGEN (1988) Estudio geotérmico del área Tuzgle-Tocomar-Pompeya. Centro Regional de Energía Geotérmica del Neuquén, pp 66. Informe Inédito. Neuquén

Giordano G, Pinton A, Cianfarra P, Baez W, Chiodi A, Viramonte J, Norini G, Groppelli G, (2013) Structural control on geothermal circulation in the Cerro Tuzgle–Tocomar geothermal volcanic area (Puna plateau, Argentina). J Volcanol Geoth Res 249:77–94

Goddart P,Wilson JJ, Couch S, Viramonte JG (1999). The evolution of El Quevar volcanic complex, Salta Province, Argentina. XIV Congreso Geológico Argentino II, pp 225–229

Golden Minerals Company (2018). El Quevar Project Salta Province, Argentina, NI 43–101 Technical Report on Updated Mineral Resource

Heidorn R (2002) Geodynamic evolution of the Central Andean back arc with respect to tertiary epithermal mineralization: examples from the Chocaya (22ºS) and La Concordia (24ºS) deposits. Dissertation zur Erlangung des Doktorgrades, Universität Salzburg, pp 111

Kay S, Coira B, Mpodozis C (1997) Southern central volcanic zone arc and back mafic magmas: signals of Andean Lithosphere processes. Actas 8º Congreso Geológico Chileno, III: 1656–1660

Kay SM, Coira BL, Caffe PJ (2010) Regional chemical diversity, crustal and mantle sources and evolution of Puna Plateau Ignimbrites in the Central Andes. J Volcanol Geoth Res 198:81–111

Kay SM, Coira B, Mpodozis C (2008) Field trip guide: Neogene evolution of the central Andean Puna plateau and southern Central Volcanic Zone. In: Kay SM, Ramos VA (eds) Field trip guides to the Backbone of the Americas in the southern and central Andes: Ridge collision, shallow subduction, and plateau uplift. The Geological Society of America Field Guide 13, pp 119–183. https://doi.org/10.1130/2008.0013(05)

Kraemer B, Adelmann D, Alten M, Schnurr W, Erpenstein K, Kiefer E, van den Bogaard P, Görler K (1999) Incorporation of the Paleogene foreland into the Neogene Puna Plateau; the Salar de Antofalla area, NW Argentina. J S Am Earth Sci 12:157–182

Marrett RA, Allmendinger RW, Alonso RN, Drake RE (1994) Late Cenozoic tectonic evolution of the Puna Plateau and adjacent foreland, northwestern Argentine Andes. J S Am Earth Sci 7:179–207

Norini G, Cogliati S, Báez W, Arnosio M, Bustos E, Viramonte JG, Groppelli G (2014). The geological and structural evolution of the Cerro Tuzgle Quaternary stratovolcano in the backarc region of the Central Andes, Argentina. J Volcanol Geotherm Res 285:214–228

Olson SF, Gilzean N (1987) Regional geology of the Nevado Queva area. Unpublished, BHP Minerals & Co, Argentina

Petrinovic IA, Mitjavilla J, Viramonte JG, Marti J, Becchio R, Arnosio M, Colombo F (1999) Geoquímica y geocronología de las secuencias neógenas de trasarco, en el extremo oriental de la cadena volcánica transversal del Quevar, noroeste de Argentina. Acta Geológica Hispana 34:255–273

Petrinovic IA, Martí J, Aguirre-Díaz GJ, Guzmán S, Geyer A, Salado Paz N (2010) The Cerro Águas Calientes caldera, NW Argentina: an example of a tectonically controlled, polygenetic collapse caldera, and its regional significance. J Volcanol Geoth Res 194:15–26

Richards JP, Ullrich T, Kerrich R (2006) The Late Miocene-Quaternary Antofalla volcanic complex, southern Puna, NW Argentina: Protracted history, diverse petrology, and economic potential. J Volcanol Geoth Res 152:197–239

Robl K, Brodtkorb M, Ametrano S (2009) La mineralización epitermal Miocena delcomplejo estratovolcánico Quevar, Salta: Parte 1: Geología y petrología. Parte 2: La mineralización. Revista de la Asociación Geológica Argentina 64(3):525–539

Sainato CM, Pomposiello MC (1997) Two dimensional magnetotelluric and gravity models of the Tuzgle volcano zone (Jujuy Province, Argentina). J S Am Earth Sci 10(3-4):247–261

Schurr B, Rietbrock A, Asch G, Kind R, Oncken O (2006). Evidence for lithospheric detachment in the central Andes from local earthquake tomography.Tectonophysics 415:203–223

Schnurr W (2001) Zur Geochemie uns Genese neogener und quartärer felsicher Vulkanite in den südlichen Zentralanden (25°–27°S und 67°–69°W): Berliner Geowissenschaftliche Abhandlungen 211:1–152

Seggiaro RE (1994) Petrología, Geoquímica y Mecanismos de Erupción del Complejo Volcánico Coranzulí. Ph.D. Thesis. Universidad Nacional de Salta, Argentina

Chapter 3
Silicic Dome Volcanic Complexes

Abstract Silicic Dome Volcanic Complexes are very well represented in Puna Norte and represent the initial manifestations of Cenozoic magmatism in the region. Their eruption, 100–200 km east of the actual frontal arc, began at the end of the Upper Oligocene–Lower Miocene and extended to the Middle Miocene in two pulses: 25–15 Ma and 13–11.8 Ma. They correspond to conical bodies with tumulus forms (0.1–4 km^2) composed by silicic lavas and shallow intrusive bodies, together or not with pyroclastic facies. They have been extruded in several episodes of Pelean to Vulcanian type with strong hydromagmatic characteristics. The domes of the first pulse are dacitic, metaluminous to weakly peraluminous, with minor andesites and trachydacytes, and the second pulse rhyodacitic–rhyolitic and peraluminous. They show LILE and LREE enrichment relative to the HREE and negative spikes in Nb, Ta, and Ti. The REE pattern slope increases from the first to the second pulse rocks (Sm/Yb = 2–3.3 to = 3.5–5) and the isotopic composition varies between$^{87}Sr/^{86}Sr$ $\sim$ 0.706 to $\sim$0.713, $\in Nd_{T\text{-}3}$ to –7 in rocks of the first cycle to the second, link with crustal thickening previous or concomitant to the second pulse and with an increase of hybridization depth of the magmas at that time. Epithermal systems (Pb-Zn) rich in Ag-Au of intermediate sulphidation are associated with the dome complexes of the second pulse.

Keywords Cenozoic · Dome complexes · Silicic · Epithermal deposits

3.1 Introduction

Silicic Dome Volcanic Complexes are very well represented in Northern Puna as the initial representatives of Cenozoic magmatism in the region, prior to the ignimbrite flare-up initiated at $\sim$10 Ma. The emission of those dome volcanic complexes, 100–200 km east of the actual frontal arc, began at the end of the Upper Oligocene–Lower Miocene and extended to the Middle Miocene. Two pulses of emission were recognized among them (Caffe 1999; Caffe and Coira 1999): Fig. 3.1 (a) Upper Oligocene–Lower Miocene represented by the dome volcanic complexes: Casa Colorada and Minuyoc (~17 Ma), Pirurayo–Cerrillos (25–20 Ma), Acay Monzonite (18.9 Ma),

B. L. L. Coira and C. E. Cisterna, *Textures, Structures and Processes of Volcanic Successions*, Springer Earth System Sciences,
https://doi.org/10.1007/978-3-030-52010-6_3

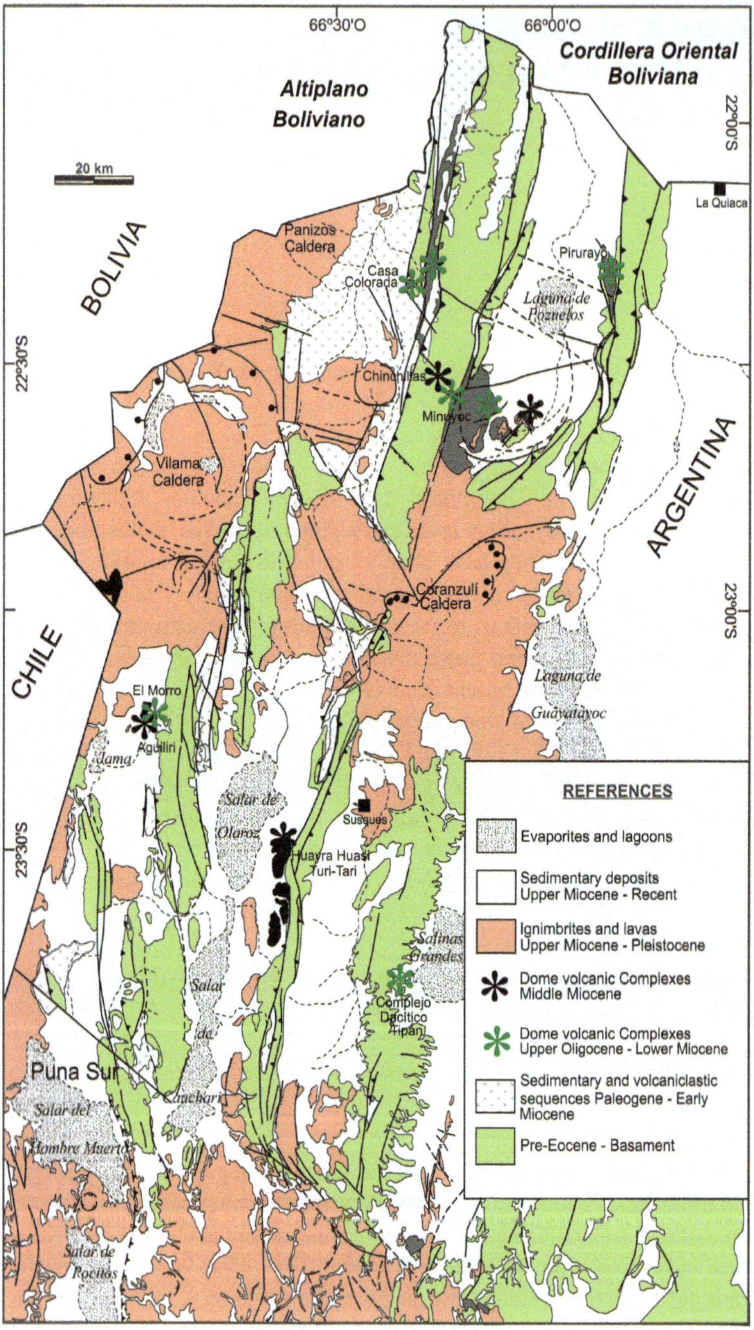

Fig. 3.1 Geologic map of Northern Puna. Asterisks in green correspond to dome volcanic complexes of the first pulse in black to the second pulse. Modified after Caffe and Coira (1999)

Tipán Dacite (15 Ma); (b) Middle Miocene dome volcanic complexes represented by Pan de Azúcar (12–13 Ma), Chinchillas (13 Ma), Cerro Redondo (12.5 Ma), Aguiliri (12.7 Ma), Huayra Huasi-Turi Tari (11.8 Ma), and Morro (12.6 Ma).

The dome volcanic complexes of Puna N are composed of one or more silicic lavic and shallow intrusive bodies of small to moderate volume with or not pyroclastic facies, whose morphology and facies allow to characterize them as "dome volcanic complexes". They show an alternation of explosive and effusive phases and minor shallow intrusive bodies. In general, they have been extruded in several eruption episodes of Pelean type alternating with others of Vulcanian style and with strong hydromagmatic characteristics. Most centers share common characteristics in their evolution such as external water that enters the magmatic system, causes fragmentation of magma, and makes way for explosive eruption; recurrence of phases of growth and destruction of lava domes which generate block and ash flows or incandescent volcanic avalanches deposits (e.g., Cerro Redondo, Casa Colorada, Chinchillas, and Pan de Azucar).

The eruptive complexity among them varies from (**a**) simple = effusive lava complexes + very shallow intrusive bodies, e.g., Aguiliri, Huayra Huasi and equivalents in Bolivia: Pulacayo, Llallagua, and San José de Oruro. (**b**) intermediate = opening + effusion, e.g., Cerro Redondo, Minuyoc, and equivalents in Bolivia: Cerro Rico, Potosí, La Joya–Kori Kollo, Todos Santos. (**c**) **Complex** = explosive and effusive episodes + construction and destruction of domes + intrusion of bodies. e.g: Pirurayo, Casa Colorada, Chinchillas, Pan de Azúcar and equivalent in Bolivia: Carangas.

During the Middle Miocene (15–12 Ma), the domic complexes reached the best conditions for greater mineralization (Caffe and Coira 1999; Caffe 2005). They resulted from the combination of several aspects: eruptive mechanisms (Peleanos and Vulcanianos) with important hydromagmatic components, siliceous magmas strongly controlled by destabilization of hydrated minerals in source areas (MASH), and with the release of fluids and metal elements. The existence of structures of both magmatic and tectonic origin, and of permeable pyroclastic facies, allowed the implantation of epithermal systems of intermediate sulphidation, with significant participation of saline fluids of magmatic origin mixed with meteoric waters (e.g., Pan de Azucar, Chinchillas, and Cerro Redondo). The magmas of the Northern Puna domes evolved at high £ O2, so Sn concentrations have been lower than similar dome complexes of the Bolivian tin belt (Caffe and Coira 1999).

From the geochemical point of view, the Dome Volcanic Complexes in Puna Jujeña are characterized by their predominant composition, dacitic rich in K2O. Although the representatives of the first pulse (17–22 ma) as the Morro stock (Mazzuoli et al. 2008) also show andesitic and trachydacytic compositions, on the contrary, it is evident that between 12 and 13 Ma the compositions riodacitic and rhyolitic are more abundant, and that andesites are restricted to mafic enclaves. From the point of view of alumina saturation, there are two distinct groups: (**a**) one to which the Morro belongs with trends metaluminous to weakly peraluminous, and indexes A/CNK that increase progressively with the increase of SiO2 (Mazzuoli et al. 2008); (**b**) the other, consisting of the rest of dacitic and rhyolitic rocks, whose A/CNK indices are higher but at the same time more scattered or poorly correlated with SiO2 (Caffe et al. 2002; Caffe and Coira 1999). The composition of major elements of the dome volcanic complexes of the Northern Puna is similar to that of other Cenozoic volcanoes of the Central Andes: Al2O3 is high (14–17%), at same as K2O (generally > 2.5%), K2O/Na2O (>1) and the sum of alkalis (5–9%). TiO2 concentrations and MgO are low (<1.5% and <4%, respectively) and those of CaO vary between 3 and 7% for andesitic and dacitic dome volcanic complexes and between 1 and 3% in the case of rhyolitic composition (Caffe and Coira 2008). The trace elements of the dome volcanic complexes also have similar concentrations to those of others more modern rocks of the Puna (see Kay et al. 1999, 2008). The incompatible elements have a higher concentration in the Middle Miocene group than the Upper Oligocene–Lower Miocene pulse. An outstanding feature observed is a steepening in the slope of the REE patterns from the rocks of the first pulse to those of the second pulse with Sm/Yb ratios from 2–3.3 in the first, to 3.5–5 in the second and La/Yb ratios 11–23 in the first and 23–34 in the second. The isotopic composition records vary from $^{87}Sr/^{8}6Sr$ ~ 0.706, $\in$ Nd $_{T\text{-}3}$ in the metaluminous rocks of the first cycle, as the Morro stock, to $^{87}Sr/^{8}6Sr \sim 0.713$, $\in$ Nd $_{T-7}$ in the second group. Caffe et al. (2002) interpreted this variation as a consequence of cortical thickening caused by compressive deformation during previous or concomitant periods to the eruption of the Middle Miocene volcanism (second pulse). These factors would have affected the depth of hybridization of primary arc magmas, which would be been greater in the eruptive cycle of 12–15 Ma than in that of 17–22 Ma.

Examples of these dome volcanic complexes will be analyzed below considering a decreasing order of eruptive complexity in their selection.

3.2 Pan de Azúcar Dome Volcanic Complex

The Pan de Azucar Dome Volcanic Complex (12±2 Ma, Coira 1999), located in the Pozuelos lagoon (22°35´S; 66°05´W), records multiple eruptive phases as Plinian and Pelean episodes and phases of the Vulcanian type (explosion breccias, surges) emitted from multiple vents linked by a ring fracture, followed by effusion of lavas and intrusions and new phases of opening, growth, and destruction of domes and ending with the construction of a dacitic dome edifice.

Caffe (2002) has recognized in the evolution of the Pan de Azucar Dome Volcanic Complex three eruptive cycles (Fig. 3.2): (**a**) first cycle: Plinian? to Pelean: violent degassing (explosive) construction–destruction of domes; (**b**) second cycle = Vulcanian–phreatomagmátic: explosive opening, interaction extrusion lavas and intrusion, hydrothermal alteration–mineralization; **d**) third cycle = Pelean: explosive opening, construction–destruction of a banded dome, and effusion of dacitic lavas (Fig. 3.3a). The first cycle begins with the collapse of the eruptive column that gives rise to co-ignimbritic breccias characterized by abundant pumice fragments accompanied by dacitic and basement lithics, in an abundant matrix (Fig. 3.3b, c). This episode

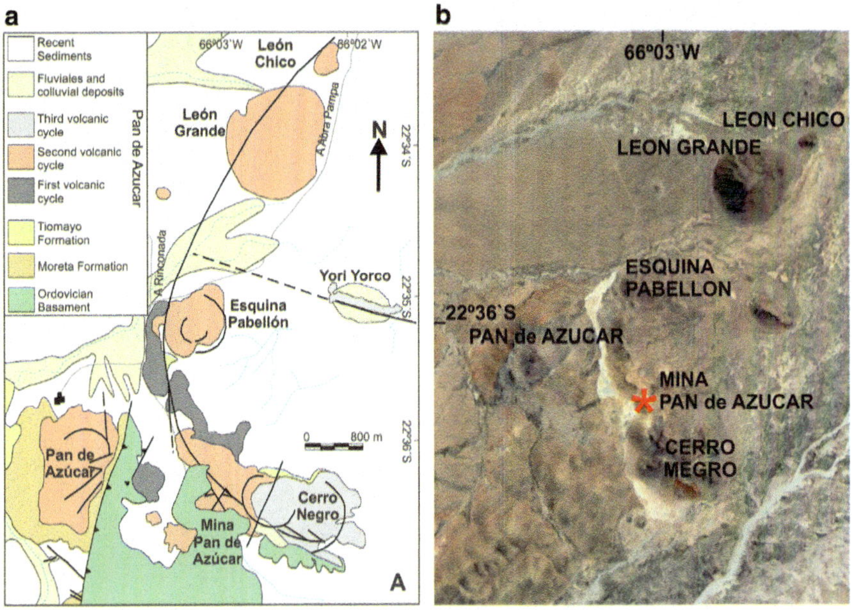

Fig. 3.2 **a** Eruptive cycles of Pan de Azucar Dome Volcanic Complex (Modified from Caffe and Coira 1999). **b** Satellite images of Pan de Azucar Dome Volcanic Complex outcrops

was succeeded by the eruption of lavas (Fig. 3.3c) and dacitic domes. The latter were associated with block and ash-flow deposits (Fig. 3.3b, d), resulting from its collapse.

The second cycle occurred mainly in the central sector, where hydromagmatic explosions—apparently directed toward the west—generated surges, and ashfall deposits (Fig. 3.4a–d).

Biotitic dacitic lavas, domes (Fig. 3.4e), and a sub-volcanic stock were later emplaced (Cerro Mina Fig. 3.4f). Hydrothermal alteration (adularia-sericita) or low sulphidation association is superimposed associated with Pb, Zn, (+Ag±Sn) mineralization (epithermal deposit rich in base metals) (Caffe 2005).

The third cycle is mainly considered a Pelean eruption, with the explosive, hydromagmatic reactivation of the system with explosion breccia (Fig. 3.5a), shifting the activity to the edge austral of the complex (Cerro Negro) along the annular fracture. The emission center was later occupied by a banded lavic dome rich in volatiles then destroyed by a final eruption of Pelean type giving place to block and ash deposits(Fig. 3.5b). Finally, magma again degassed, erupted as dacitic lavas with biotite and hornblende (Fig. 3.5c–e).

In the geological interpretation of the aeromagnetic survey of the Northern Puna in the area of the depression of Pozuelos, Coira et al. (1996) proposed a buried calderic structure located in it, which included the Pan de Azucar Dome Volcanic Complex. Modeling on magnetic anomalies detected in a ground magnetic sounding (Prezzi 2002) and three-dimensional modeling of the magnetic data supported the existence

Fig. 3.3 First Cycle: **a** Co-ignimbrite breccia, **b** Satellite image with the distribution of first h cycle volcanic facies CI (Co-ignimbrite breccia); L (Dacite lava flows); BA (Block and ash deposits). **c** Dacite Lava flows, **d** Block and ash deposits. Blocks with homogeneous dacitic composition show "bread crust structures" and cooling fractures

of an intrusive body (1 km in diameter at approximate depths of 250 m) in it (Prezzi and Gótze 2006), and new detailed ground magnetic and gravity surveys (Prezzi and Lince Klinger 2010) supported the hypothesis of a large caldera buried beneath the Laguna de Pozuelos Basin. All these results obtained supporting the hypothesis of a large caldera buried beneath the Laguna de Pozuelos Basin which host the Pan de Azucar Dome Volcanic Complex and other intrusive bodies are valuable for further mining exploration in the area considering the Pb, Zn, (+Ag±Sn) mineralization associated with that Dome Volcanic complex.

3.3 Chinchillas Dome Volcanic Complex

The Chinchillas Dome Volcanic Complex (13 ± 1 Ma), of less eruptive complexity than Pan de Azucar, is located in the Sierra de Rinconada (22°30′S; 66°15′W) hosted in Ordovician sedimentary rocks. It is constituted by breccias, blocks and ash, ignimbrites and ashfall tuffs and surge deposits that fill a depression or diatreme volcanic throat of elliptical shape, approximately two kilometres long by 1.6 kms wide, associated with collapses in ring fractures, and infilled with the pyroclastic rocks (Coira et al. 1993; Caffe et al. 2008). At the contact between pyroclastic

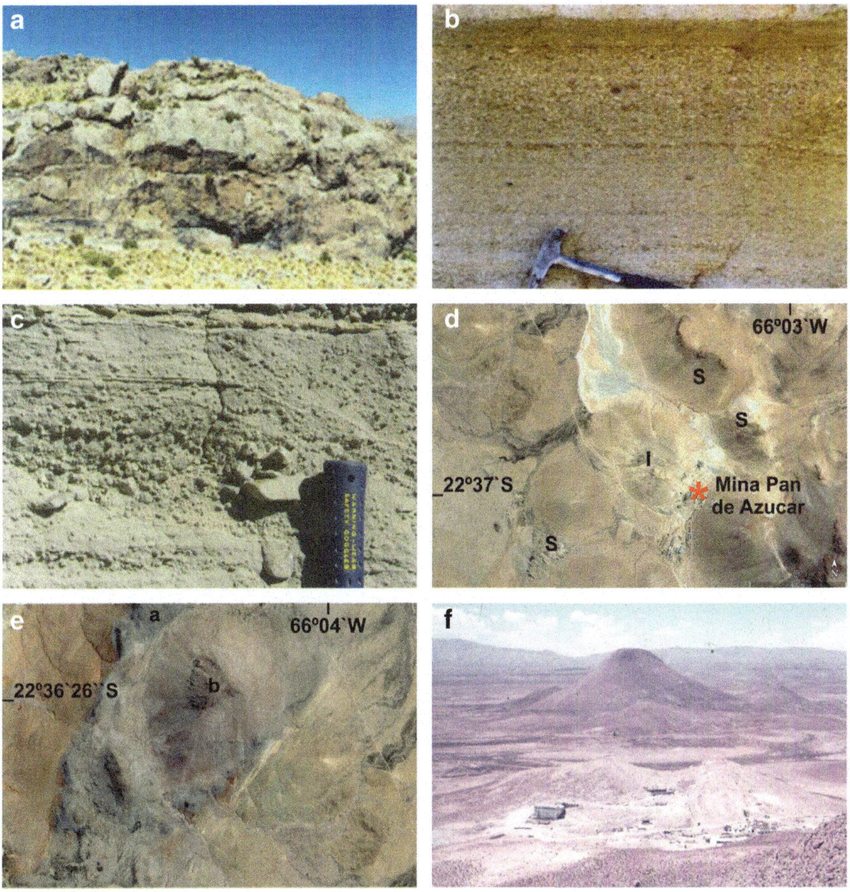

Fig. 3.4 **Second Cycle**: **a** outcrop of surge and ashfall deposits. **b** surge deposits with laminar structure. **c** accretionary lapilli associated with phreatomagmatic eruption. **d** satellite images with the distribution of second cycle volcanic facies around Pan de Azucar Mine: S (surge and ashfall deposits); I (Cerro Mina Dacite Intrusive hydrothermal alterated and mineralized). **e** Cerro Pan de Azucar Dacite lava flows (**a**), dome autobreccias of dacitic composition (**b**). **f** Pan de Azucar Mine. Dacite sub-volcanic stock of Cerro Mina hydrothermal alterated and mineralized. Cerro Pan de Azucar in the background

volcanics and Ordovician Basement metasediments, a wide zone of hydraulic fracturing and brecciation of the basement has formed. Dacitic lavas, flow domes, and sub-volcanic intrusions occur on the southern margin of the basin at the contact between metasediments and pyroclastics (Figs. 3.6, 3.7, 3.8).

The first event in the evolution of the complex was an annular collapse as a result of explosions by magmatic volatiles, which has produced a wide zone of hydraulic fracturing and brecciation of the Ordovician Basement with the formation of a collapse breccia (Fig. 3.9a, b). This collapse breccia emplaced along the annular fracture zone

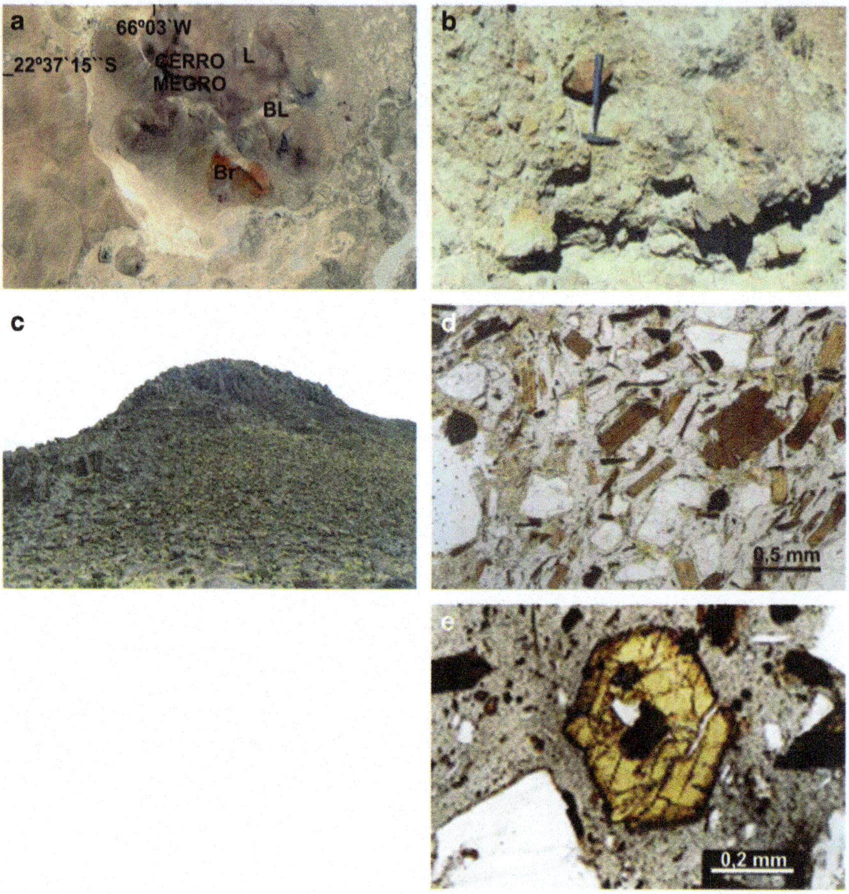

Fig. 3.5 **a** Satellite images with the distribution of third cycle volcanic facies, in the Cerro Negro sector. Br (explosion breccia), BL (block and ash deposits), L (dacite lava flows). **Third Cycle**: **b** Block and ash deposit showing chaotic arrangement, cognitive dacitic clasts with bread crust rims in ashy matrix. **c** Dacite lava flows as the posthumous event. **d** In thin section, porphyric dacitic lava flow·30–40% phenocryst of plagioclase, quartz, biotite, hornblende, and as accessories±cpx apatite, titanomagnetite, ilmenite Plane polarized light (courtesy of P. Caffe). **e** In thin section, detail of the dacite lava flow with hornblende. Plane polarized light (courtesy of P. Caffe)

is composed fundamentally by angular clasts of the Ordovician Basement (2–40 cm in diameter) with jigsaw-fit texture (Fig. 3.9c, d) cemented with argentiferous galena plus a stock working of sphalerite–siderite–galena which in places carries low-grade silver–zinc mineralization along widths of over 300 m.

Afterwards, an early explosive phase represented by pyroclastic breccias, ignimbrites, and tuffs (Fig. 3.10a–c) erupted from along the annular fracture that delimited the depression in the south and filled the resulting depression, contouring the vent walls, dipping moderately toward the center.

Fig. 3.6 Satellite images of Chinchillas Dome Volcanic Complex. Volcaniclastic deposits filling the depression in yellow, lava domes in green. Dashed line corresponds to collapse ring fracture

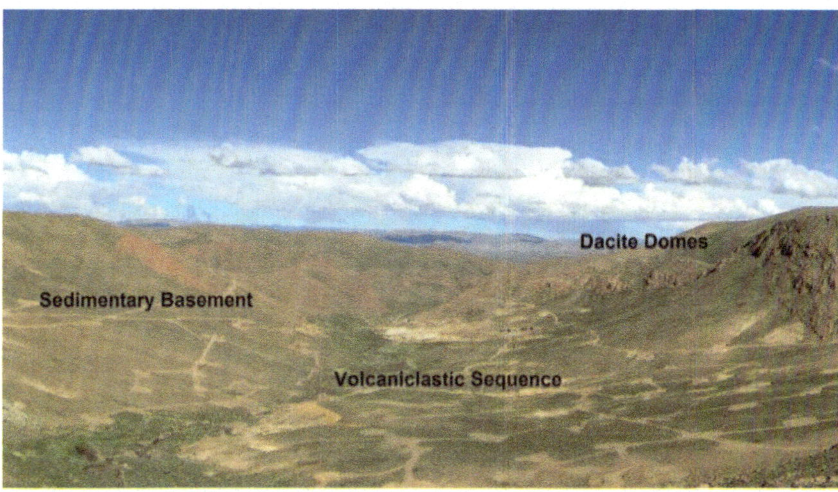

Fig. 3.7 View of the Chinchillas deposit, looking east showing outcrop of the sedimentary Ordovician Basement rocks, the volcaniclastic sequence infilling the depression, and the dacite lava domes flanking the southern border of the deposit

The breccias are mainly matrix-supported but sometimes clast-supported. The clasts predominantly reworked pyroclastic tuffs, lava fragments, and intrusive fragments of dacitic composition are subrounded to angular and vary from fine-grained to large metre-scale blocks. The ignimbrites and tuffs (Fig. 3.10b, c) are composed by crystalloclasts (20–25%) of quartz, andesine, biotite and as accessory apatite, zircon,

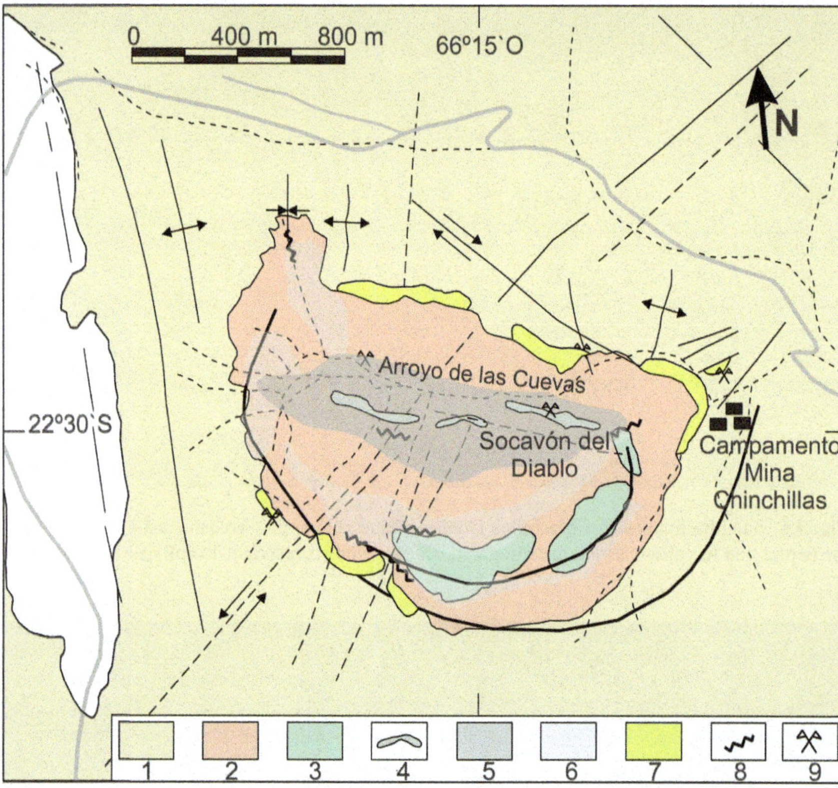

Fig. 3.8 Geological mining and hydrothermal alteration map of the Chinchillas Dome Volcanic Complex, modified from Coira et al. (1993), Caffe and Coira (1999) (1) Ordovician Basement; (2) Ignimbrites, breccias, and ashfall tuff, block and ash deposit; (3) Dacite lava domes; (4) Explosiбn breccia, ore carrier, silicified and with replacement by feldspar; (5) Steam-heated zone (sericite-kaolinite) superimposed on feldspatization; (6) Post-feldspatization sericitization–argilitization zone; (7) Collapse breccias of the Basement with ore; (8) Late Fe–Mn deposits; (9) Old mining workings

and ilmenite. Pumice fragments (10–15%) and scarce Ordovician sediments and dacite lithics are distributed in a vitroclastic matrix altered by intense hydrothermal activity.

Subsequently, the extrusion of dacite domic lavas and minor episodes of collapse of the growing domes took place along the annular fracture that delimits depression in the south (Fig. 3.11a). The domes have a medium to fine-grained porphyritic texture with phenocrysts of quartz, (35–45%) plagioclase, biotite, and minor sanidine (Caffe and Coira 2008).

The dacite domes are generally massive in texture with limited flow banding and some flow brecciation along the margins. They lie horizontally above tuff breccias. Drilling confirms that the dacite outcrops are part of larger bodies located below the Socavon del Diablo area (Puna Operations Inc 2017).

Fig. 3.9 **a** Ordovician Basement affected by annular fracture with the development of collapse breccia hydrothermalized and mineralized. **b** Detail of old Chinchillas mine hosted in the zone of mineralized collapse breccia. **c** Brecciated Ordovician Basement sediments with fine volcanic matrix near the contact between pyroclastic sequence and the Basement. **d** Contact between collapse breccia and sedimentary Ordovician Basement

The block and ash deposits (Fig. 3.11b), the result of the collapse of growing domes, are composed of a large fraction of juvenile dacitic blocks in a matrix of volcanic ash of the same composition. Some of the blocks contain radially arranged cooling joints which show they were emplaced as hot blocks.

Mineralization (Ag Zn Pb Sn)± is distributed in four main areas: the Silver Mantos and Mantos Basement zones in the western part of the deposit, and the Socavon del Diablo (Fig. 3.12a–c) and Socavon Basement zones in the eastern part (Puna Operations Inc 2017). The main structural elements controlling the location of mineralization are the inverted cone shape of the volcanic diatreme forming the contact between basement sediments and overlying volcanic rocks; and the dominant east–west and

Fig. 3.10 **a** Overview looking south of hydrothermalized volcaniclastic sequence in the central depression. In the background, the dacite domes flanking the southern border of the complex. **b** Outcrop of massive ignimbrite on top of ash-tuff and surge deposits hydrothermalized. **c** Outcrops of ignimbrite beds with contrasting induration and hydrothermal alteration

Fig. 3.11 **a** Dacitic lava domes outcropping along the southeast edge of the Chinchillas Dome Volcanic Complex. **b** Dacite block and ash deposit on top of pyroclastic breccias. At the base of the depression, the volcaniclastic deposit highly hydrothermalized (courtesy of P. Caffe)

subordinate north-west, north, and north-northeast trending structures that control the formation of the Chinchillas volcanic center.

The phreatomagmatic explosion that produced the diatreme generated a symmetrical cylindrical-shaped caldera, with mineralized brecciated basement rocks along the contacts and horizontal layers in the tuffs with disseminated mineralization.

Fig. 3.12 **a** Volcaniclastic deposit highly hydrothermalized in the area of the Socavon del Diablo. **b** Coarse-grained clast-supported breccia at Socavon del Diablo. Mineralization infilling matrix and open spaces. **c** Outcropping of explosion breccias at Socavon del Diablo

3.4 Cerro Redondo Dome Volcanic Complex

The Cerro Redondo Dome Volcanic Complex (12, 54 ± 1, 1 Ma, Cladouhos et al. 1994) (Fig. 3.13a, b) is hosted in Ordovician sedimentary rocks in the Sierra de Rinconada (22°22´S; 66°08´W). It is of intermediate eruptive complexity, composed by basal pyroclastic deposits (hydromagmatic explosion breccias) which represent the opening stage, the result of external water that enters the magmatic system and causes fragmentation of magma and makes way for explosive eruption. These deposits were succeeded by the effusion of dacitic lavas that formed the Cerro Redondo lava dome (Fig. 3.14c) extending south. Faults of WNW-ESE and N-S strike control the location of the dome complex, while nested annular fractures delimit possible areas of collapse.

Pyroclastic deposits up to 60 m thick are distributed as a ring outcropping preferentially to the north, while to the south predominate the lavas, forming an asymmetric dome complex. The exposed level of the complex is of relative depth observed in basal areas of the dome feeding ducts. The basal pyroclastic levels correspond to hydromagmatic explosion breccias, with mainly dacitic clasts up to 30 cm in diameter and accidental of the Ordovician basement (Fig. 3.14b). These gradually continue to higher finer levels rich in matrix with clasts < 15 cm in diameter, massive or with planar stratification (Fig. 3.14d).

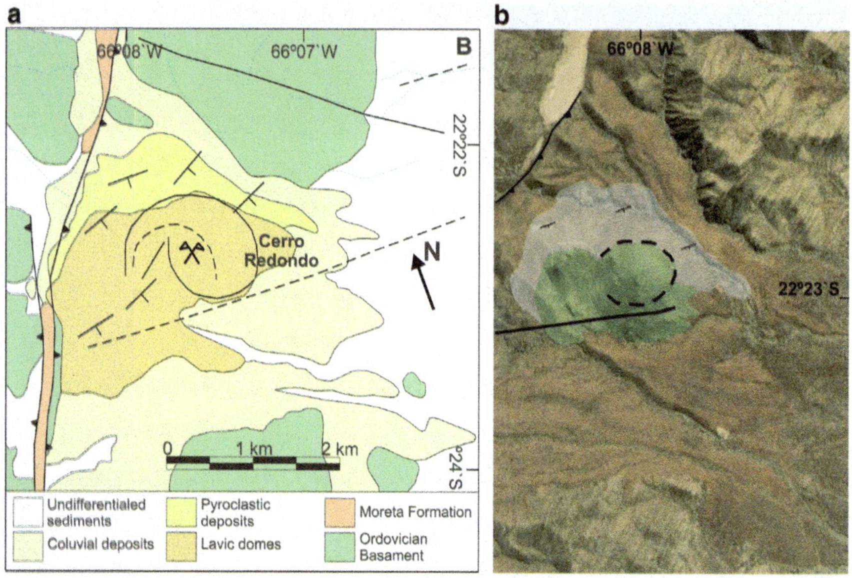

Fig. 3.13 **a** Map of Cerro Redondo Dome Volcanic Complex (CRDVC); modified from Caffe and Coira (2008). **b** Satellite images of (CRDVC) dacitic lavas in green and pyroclastic deposits in gray, annular fracture and fault in dash line

The dacitic lavas, frequently hydrothermal alterated, have a porphyric texture and are formed by plagioclase, quartz, biotite and apatite and zircon as accessories, distributed in a vitreous to micro-grain paste (Caffe 2005).

The Dome Complex is hydrothermal altered and mineralized. The central zone of the domic structure is brecciated, silicified with tourmaline along the annular fractures that affect the complex, as well as the numerous radial structures that generated during the collapse of the dome building (Caffe 2005). Surrounding that alteration zone there is a feldspatization zone replaced by sericitization and argilitization. Propylitic halo altered mainly the pyroclastic deposits. The ore was deposited jointly to the brecciation and to hydrothermal alteration represented by tourmaline and silicification. Brito and Sureda (1992) define a series of hypogenic metalliferous mineralization pulses, which deposit Pb, Zn, Cu, As, Sb, Fe, Au, and Ag sulfides and sulfosalts.

3.5 Aguiliri Volcanic Complex

The Aguiliri Volcanic Complex (23°11′S-23°17′S; 66°51′W-66° 57′W) includes a cluster of two stocks: dacitic (Aguiliri) and andesitic (Morro) and dacitic lava domes (La Vega Chingolo, Norte and Sin Nombre hills which intruded and overlayed the Ordovician Basement and Tertiary sediments) (Fig. 3.15). They outcrop, controlled by structural lineaments WNW-ESE and N-S (Gorustovich et al. 1985).

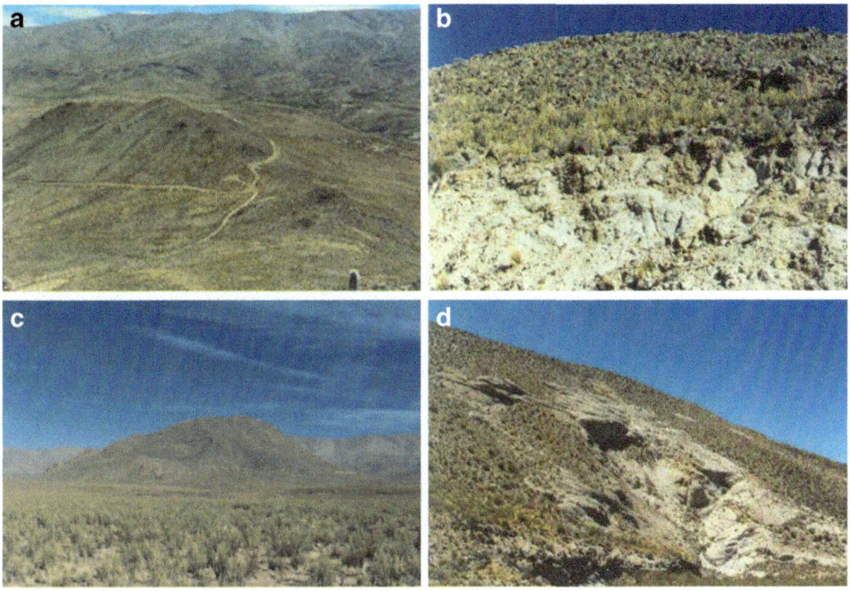

Fig. 3.14 **a** Overview looking west of Cerro Redondo Dome Volcanic Complex. Pyroclastic deposits (on the right) underlying and surrounding the dacitic lavas of Cerro Redondo in the form of an “apron”, **b** Hydromagmatic explosion breccias, with mainly dacitic clasts up to 30 cm in diameter and accidental of the Ordovician basement, underlying dacitic lavas, **c** Dacite lava dome of Cerro Redondo looking southwest. **d** High levels of massive fine breccias with up to 15% accidental fragments, with evidence of water interaction

With the exception of Cerro Moro stock of quartz andesitic composition, the other centers have phenodacitic composition, with abundant plagioclase, quartz, biotite, and hornblende (except in the Aguiliri stock, where the last is absent). Allanita, apatita, zircon, monazite, and opaque are minor accessories. The Aguiliri stock (12.7 ± 1.3 Ma; Caffe et al. 2002), intrusive in the Tertiary sedimentary sequence, has anomalously high concentrations of U and weak anomalies of Pb and Ag along a hydrothermal and radioactive fault belt that affects intrusive and sedimentary host rock. Aniel et al. (1992) concluded that the U enrichment was hydrothermal not magmatic. The Morro, a small quartz andesitic stock which intrudes the Ordovician Basement, shows a higher degree of erosion than the other domes and has closer geochemical and mineralogical affinities with the dome complexes of the Upper Oligocene–Lower Miocene pulse than the Middle Miocene pulse represented by the rest of the centers of the Aguiliri Volcanic Complex (Caffe et al. 2002).

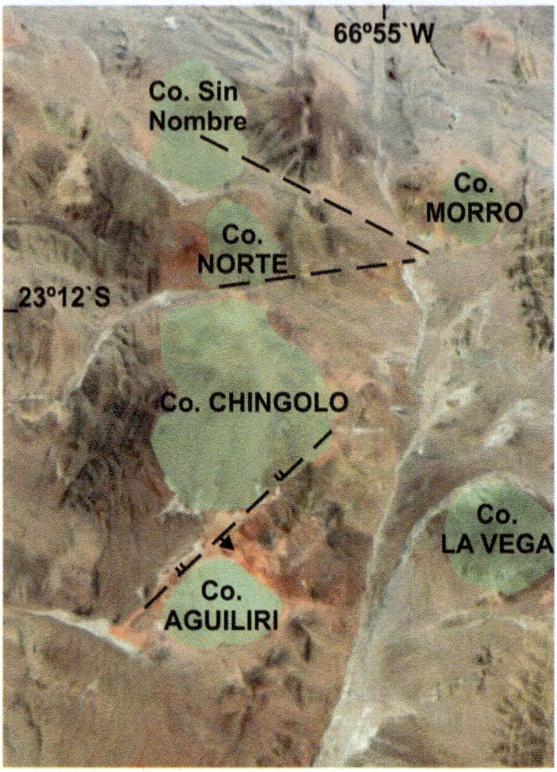

Fig. 3.15 Satellite images of Aguiliri Volcanic Complex

References

Aniel B, Vullien AR, Romano HI (1992) Estudio genético del yacimiento uranífero Aguiliri-Puna Argentina. Rev de la Asociación Geol Argentina XLIV (1–4):3–14(1989)

Brito J, Sureda R (1992) Exploración del Prospecto polimetálico Cerro Redondo, Rinconada, Jujuy, Argentina. 4ª Congreso Nacional de Geología Económica, Córdoba, pp 183–194

Caffe PJ (1999) Complejos volcánicos dómicos del terciario superior en Puna Norte: sus implicancias magmatotectónicas y metalogenéticas. Universidad Nacional de Cordoba, facultad de ciencias Exactas, Físicas y naturales Tesis doctoral, inédita 421 páginas

Caffe PJ (2002) Estilos eruptivos del Complejo volcánico dómico Pan de Azúcar-Puna Norte. Rev de la Asociación Geol Argentina 57(3):232–250

Caffe PJ (2005) Alteraciones hidrotermales asociadas a áreas mineralizadas del NOA. Complejos volcánicos dómicos de Pan de Azúcar, Chinchillas y Cerro Redondo. Provincia de Jujuy. Serie Contribuciones Técnicas-Recursos Minerales 24:1–75

Caffe PJ, Coira BL (1999) Complejos de domos volcánicos del Mioceno medio de Puna Norte. Un modelo geológico y metalogenético para yacimientos epitermales de metales de base ricos en plata (estaño). In: Zappettini, E.O. (Ed.): Recursos Minerales de la República Argentina. Instituto de Geología y Recursos Minerales-SEGEMAR. Anales 35:1569–1578. Buenos Aires

Caffe PJ, Coira B (2008) Depósitos epitermales polimetálicos asociados a complejos volcánicos dómicos: Casa Colorada, Pan de Azúcar, Chinchillas y Cerro Redondo. In: Coira B, Zappettini EO (eds) Geología y Recursos Naturales de la Provincia de Jujuy, Relatorio del XVII Congreso Geológico Argentino. Asociación Geológica Argentina. Buenos Aires, pp 350–357

Caffe PJ, Trumbull RB, Coira B, Romer R (2002) Petrogenesis of early Neogene Magmatism in the Northern Puna. Implications for magma genesis and crustal processes in the Central Andean Plateau. J Petrol 43(5):907–942

Caffe PJ, Coira B, Kay SM (2008) Magmatismo neógeno temprano de la Puna jujeña. In: Coira B, Zappettini EO (eds) Geología y Recursos Naturales de la Provincia de Jujuy, Relatorio del XVII Congreso Geológico Argentino. Asociación Geológica Argentina. Buenos Aires, pp 302–312

Cladouhos TT, Allmendinguer RW, Coira B, Ferrer E (1994) Late Cenozoic deformation in the Central Andes: fault kinematics from the northerm Puna, northwestern Argentina and southwestern Bolivia. J S Am Earth Sci 7(2):209–228

Coira B, Díaz A, Chayle W, Pérez A, Ramírez A (1993) Chinchillas, un modelo de complejo volcánico dómico portador de depósitos de metales de base con Ag y Sn en Puna jujeña. In: Actas XII Congreso Geológico Argentino-II Congreso de Exploración de Hidrocarburos, vol IV, pp 270–276

Coira B, Caffe P, Chayle W, Rodríguez G, Pérez A, Soler M, Rankin L (1996) Interpretación geológica del relevamiento aeromagnético de la Puna Septentrional, Jujuy y Salta. II Area Depresión de Pozuelos. II. I Geología y Minería. Serie Contribuciones Técnicas (1) Geofísica, pp 1–13. Dirección Nacional del Servicio Geológico

Gorustovich SA, Salfity J, Rubiolo DG, Sureda RJ (1985) Yacimiento Aguiliri (Puna, Argentina) Depósito uranífero en rocas subvolcánicasterciarias IV Congreso Geológico ChilenoII83). Antofagasta, pp 250–270

Kay S, Mpodozis C, Coira B (1999) Neogene Magmatism, tectonism and mineral deposits of the Central Andes (22°-33°S latitude). In Skinner BJ (ed) Geology and ore deposits of the Central Andes. Society of Economic Geology Special Publication, vol 7, pp 27–59

Kay SM, Coira B, Caffe PJ (2008) Geoquímica, fuentes y evolución de magmatismo neógeno de la Puna Norte. In: Coira B, Zappettini EO (eds) Geología y Recursos Naturales de la Provincia de Jujuy, Relatorio del XVII Congreso Geológico Argentino. Asociación Geológica Argentina. Buenos Aires, pp. 322–334

Mazzuoli R, Vezzoli L, Omarini R, Acocella V, Gioncada A, Matteini M, Dini A, Guillou H, Hauser N, Uttini A, Scaillet S (2008) Miocene magmatism and tectonics in the easternmost sector of the Calama-Olacapato-El Toro fault system in Central Andes at~24°S: insights into the Eastern Cordillera evolution. Geol Soc Am Bull 120(11/12):1493–1517

Prezzi C (2002) Relevamiento magnetométrico terrestre de la zona de Pan de Azúcar, Puna Jujeña (22°30'S - 66°00'O). Rev de la Asociación Geol Argentina 57(1):85–91

Prezzi C, Götze H-J (2006) 3D modeling of buried intrusives in Pan de Azúcar zone (northern Puna, Argentina) from ground magnetic data. J S Am Earth Sci 22(1–2):89–97

Prezzi C, Lince Klinger F (2010) Nuevas evidencias geofísicas de la existencia de una caldera en Laguna Pozuelos, Puna Norte. Rev de la Asociación Geol Argentina 66(1):280–293

Puna Operations Inc. (2017) NI 43-101 Technical Report Pre-feasibility Study of the Chinchillas Silver-Lead-Zinc Project Jujuy Province, Argentina, p 231

Chapter 4
Monogenetic to Polygenetic Mafic Volcanism

Abstract The mafic back-arc volcanism, uncommon in the Andean Central Volcanic Zone, is represented in the Puna region by late Miocene to Quaternary small monogenetic to polygenetic cinder cones and lava flows, as isolated or poorly clustered centers of basaltic andesitic to andesitic composition. Maars, tuff rings, and domes are also usual. This volcanism has a reduced representation in relation to the voluminous silicic ignimbrites and andesitic–dacitic stratovolcanoes that characterize the region. The scoria cones have moderate to low maximum slopes <20° and are <100 m high and 300–1300 m in diameter, common subcircular. Most lava flows were erupted from boccas at the scoria cone base or from the crater and reach length up to 3 km. These centers are concentrated along distensive and strike-slip faults (NW–SE, E–W, NE–SW) that favored magma ascent. Their facies and architecture are consistent with eruptions mainly Strombolian, with brief periods of lava fountaining and hydrovolcanism, the latter usually restricted to the early stages of cone construction. In Northern Puna this volcanism, Late Miocene–Early Pliocene, is represented by basaltic andesites to andesites, high-K calc-alkaline with minor shoshonites. In Southern Puna this volcanism, active till Holocene, is represented by basalts and basaltic trachyandesites belonging in equal measure to calc-alkaline and within-plate groups.

Keywords Cenozoic · Puna · Mafic volcanism

4.1 Introduction

The mafic back-arc volcanism, uncommon in the Central Volcanic Zone, is represented in the Puna region by late Miocene to Quaternary small monogenetic to polygenetic cinder cones and lava flows, as isolated or poorly clustered centers of basaltic andesitic to andesitic composition. Maars, tuff rings, and domes are also usual. Although this back-arc mafic volcanism has a smaller representation with smaller volumes in relation to the calderas and stratovolcanoes in the Andes, they are of particular significance since they offer insights into the state of the mantle under the world's second-largest continental plateau. Their scoria cones have moderate to low

B. L. L. Coira and C. E. Cisterna, *Textures, Structures and Processes of Volcanic Successions*, Springer Earth System Sciences,
https://doi.org/10.1007/978-3-030-52010-6_4

maximum slopes <20° and are <100 m high and 300–1300 m in diameter, common subcircular, some with aprons of ash and lapilli deposits scattered with larger clast. Most lava flows were erupted from boccas at the scoria cone base or from the crater and reach length up to 3 km. Their facies and architecture are consistent with eruptions mainly Strombolian, with brief periods of lava fountaining and hydrovolcanism, the latter usually restricted to the early stages of cone construction. Changes in eruptive style are thought to be caused by variations in both the internal (e.g., magma ascent) and external (e.g., surficial water availability) conditions (Maro and Caffe 2016). The absence of hiatus between or within the different lithofacies, observed in the Puna Norte centers and in several of the Southern Puna, conforms to the definition of "monogenetic volcano" sensu stricto: *volcanic building of small volume (≤1 km^3) built from a continuous eruption, or several discontinuous ones developed in a short time* (Németh and Kereszturi 2015). Some volcanoes in the Southern Puna exhibit compositional variations (polymagmatism, in the sense of Brenna et al. 2010) and also interruptions in the eruptive activity of even thousands of years corresponding to the "polycyclic monogenetic volcanoes" (Németh and Kereszturi 2015).

It is worth noting that almost all the mafic centers in the Puna show a relatively low degree of degradation compared with the Miocene–Pliocene monogenetic cones in other regions of the world (Hooper and Sheridan 1998; Kereszturi and Nemeth 2012) in agreement with the high aridity conditions that have prevailed in the Puna region at that times.

Strong link between tectonic structures and the emplacement of monogenetic volcanoes was observed in the Puna. These centers are concentrated along distensive and strike-slip faults, orogeny-transversal (NW–SE, E–W, NE–SW) suggesting that they have been favored magma ascent (Marrett et al. 1994; Riller and Oncken 2003; Petrinovic et al. 2006; Montero López 2009).

In Northern Puna this volcanism, Late Miocene–Early Pliocene, is represented by basaltic andesites to andesites, high-K calc-alkaline with minor shoshonites. Meanwhile in Southern Puna, active till Holocene, it also records basalt and basaltic trachyandesites, and calc-alkaline and within-plate groups are about equally represented.

This chapter will analyze field characteristics of such volcanic centers, geometry, structure, texture, lithology, and compositional characteristics through selected examples from Northern and Southern Puna.

4.2 Northern Puna Mafic Volcanism

The Neogene mafic volcanism of the Northern Puna is represented by scoria cones and lava flows as isolated or poorly clustered centers or forms small mafic volcanic fields of up to c. 100–200 km^2 (Fig. 4.1). The scoria cones have moderate to low maximum slopes <20° and are <100 m high and 300–1300 m in diameter (Maro and Caffe 2016).

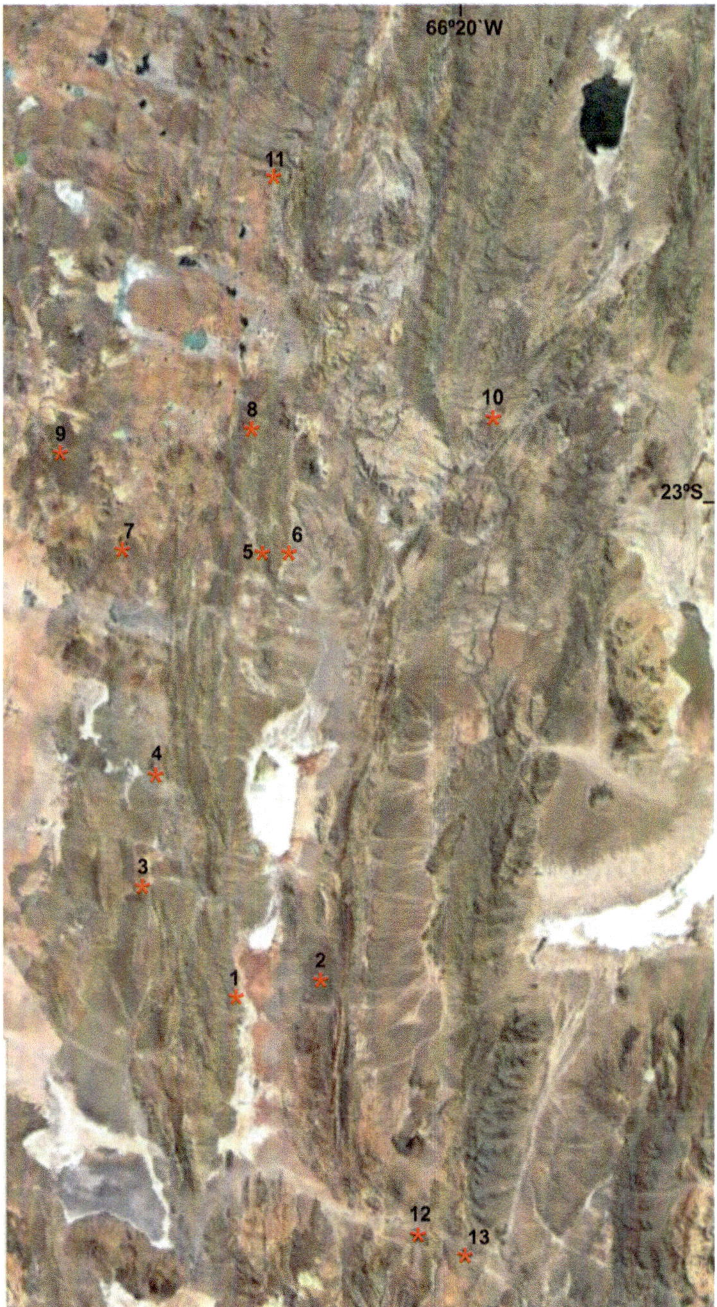
66º20`W
23ºS
11
10
8
9
7
5
6
4
3
2
1
12
13

◂**Fig. 4.1** Distribution of Late Miocene to Early Pliocene mafic volcanic centers of the Northern Puna (modified from Caffe et al. 2002). Isolated centres: (1) Cauchari, (11) Pabellón mine, (3) Cerro Tropapete, (10) Cerro Barro Negro, 7) Patahuasi, (12) San Gerónimo, (13) Chorrillos; Poorly clustered: (4) Cerros Negros de Jama, (5) El Toro, (6) Campo Negro; Small mafic volcanic fields: (8) Cerro Morado, (9) Cerro Bitiche, (2) Cerro Negro de Olaroz

They are commonly subcircular in plan view, although some elliptical shapes are seen in cones in intrinsic relationships with transverse lineaments. They are frequently aligned in a NS to NE–SW direction, in coincidence with the orientation of major thrusts in this segment of the Andes. These observations indicate a strong link between tectonic structures and the emplacement of monogenetic volcanoes in the Northern Puna. This spatial relation suggests, as proposed by several researchers (e.g., Marrett and Emermann 1992; Riller and Oncken 2003; Guzmán et al. 2006; Petrinovic et al. 2006; Norini et al. 2013), that the Neogene volcanism and compression occurred at the same time, with the regional stress field controlling the migration and emplacement of magmas. Thrust faults acted as magma channels at least in their trajectory to the surface—possibly favored by transcurrent components (Riller and Oncken 2003).

The age of mafic volcanism is restricted to the Late Miocene–Early Pliocene, slightly older than the peak of activity of the monogenetic centers in the Southern Puna region and contemporaneous with eruptions of some of the most important silicic calderas of the Altiplano–Puna Volcanic Complex.

The volcanic activity of those mafic monogenetic cones was mainly Strombolian, with occasional changes to transitional Hawaiian eruptions when the ascent speed and deposition rate were higher. The effusion of lavas was also ubiquitous and mostly coeval with explosive volcanism. Hydrovolcanic events recorded are considered to be minimal in volume and frequency and were always related to the early phases of the volcanic activity of cones. (Maro and Caffe 2016).

The absence in those scoria cones of unconformities between or inside, different lithofacies and clear evidences of interruptions in deposition in them, together with transitional passage among different lithofacies, suggests a rapid and continuous evolution consistent with the monogenetic character of this volcanism.

Based on the field characteristics of such volcanic products in Northern Puna, as geometry, structure, texture and lithology, Presta and Caffe (2014); Maro and Caffe (2016); Maro et al. (2017) have organized them into ten principal lithofacies comprising fall pyroclastic and ballistic material with variable grain size and degree of agglutination, lava flows and dykes, representing deposits from Strombolian or Hawaiian activity during the construction of scoria cones and as result of hydromagmatic processes. The predominance of lithofacies of non-welded bombs and lapilli beds in the architecture of edifices was interpreted (Maro and Caffe 2016) as the result of a strong contribution from Strombolian-type activity. Hawaiian-style eruptions, recorded by the deposition of lithofacies of spatter deposits of moderately to highly agglutinated juvenile volcanic fragments, were short and occasional, with lower

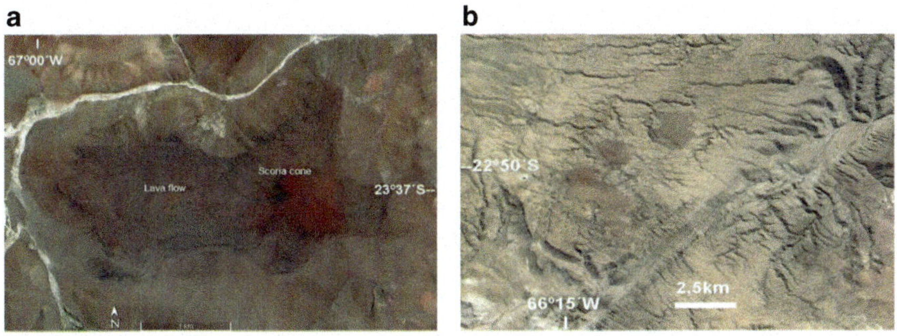

Fig. 4.2 **a** Cerro Tropapete an Upper Miocene mafic monogenetic center. Lava flow ~3 km long, aligned W–E covered by a scoria cone. **b** Cerro Barro Negro with shoshonitic lavas >4 km long and a small cone of scoria of 50 m high and remnants of a major flow

volumes of those deposits. Otherwise, deposits that suggest magma–water interactions (hydrovolcanism) were mostly found at the base of some scoria cone remnants (e.g., Cerro Bitiche, Cerro Morado, the El Toro region, and Cerros Negros de Jama). Lava flows are the most widely distributed products in the Northern Puna monogenetic volcanoes, most of them erupted from the base of cones and later covered by explosive eruptions (e.g., the Cerros Negros de Jama, Cerro Barro Negro, Cerro Tropapete). Maars and tuff rings are absent in the Northern Puna, which is consistent with the arid climate that has dominated the region at that times.

Among the mafic centers distributed in Northern Puna, Maro and Caffe (2016) have recognized as isolates mafic centers: Cerro Tropapete and Cerro Barro Negro, Cauchari, Pabellón mine centers (see Fig. 4.1). As an example of this type of centers, the Cerro Tropapete (7.2 ± 0.4 Ma, Schwab and Lippolt 1974) is characterized by a lava flow of ~3 km long, aligned W–E, later covered by a scoria cone. In this case erosion masks the classification of the type of flow because surfaces are not well preserved (Fig. 4.2a).

Another case is that of Barro Negro which overlay on 6,6 Ma ignimbrites (Fig. 4.1b.3). It consists of a small cone of scoria (50 m high) that presents a lava flow of 4.2 km in length directed toward the SSE. In the NE also appear two isolated and equidimensional lava fields that would be remnants of a major flow (Fig). This center constitutes the first record of shoshonitic lavas in Northern Puna (Maro and Caffe 2013).

As poorly clustered centers have been recognized: Cerros Negros de Jama and El Toro-Campo Negro by Maro and Caffe (2016).

The Cerros Negros of Jama (Fig. 4.3a) studied by Maro and Caffe (2012) is integrated by four centers which consist of scoria cones and lava fields which show subregional alignment coinciding with subparallel thrust fault which limits the Ordovician basement block. The lava flows associated with the occidental cones extend mainly toward the NE and were extruded from the base of the cones and would have been fed by dykes. In the case of the oriental cone, lava with different lengths flowed toward NW and SE. It is observed in the different cases over-lapping of several lava flows,

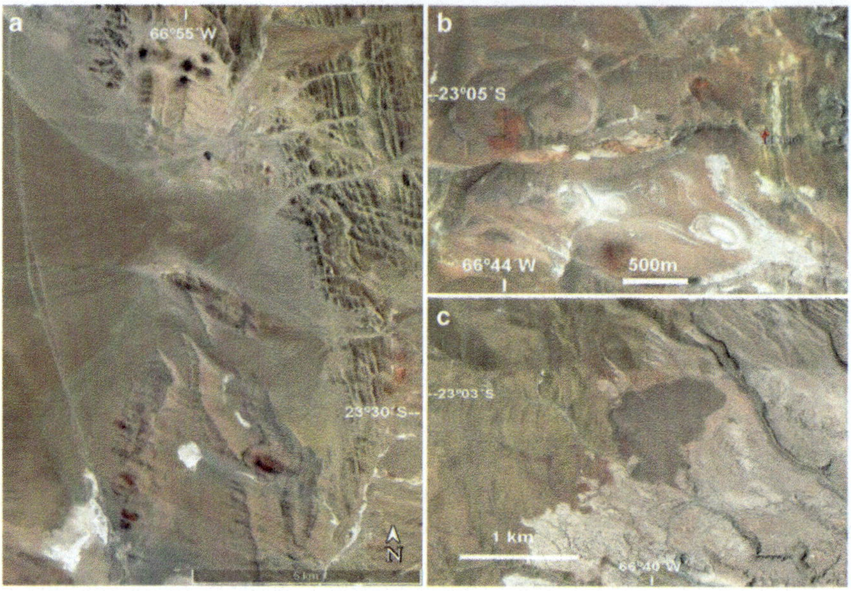

Fig. 4.3 **a** Cerros Negros de Jama integrated by four centers which consist of scoria cones and lava fields which show subregional alignment. **b** El Toro. A cluster of cinders cones with associated andesitic to basaltic andesites lava that flowed mostly to the east and southeast. **c** Campo Negro monogenetic center composed of a small cinder cone and a succession of andesitic to basaltic andesitic lava flows foliated by flow with ENE direction

those that show variable lengths (900 m–3 km). As part of this set of centers, and the most north of them was composed exclusively by a succession of lava flows, without identifiable center.

On the other hand in the monogenetic volcanoes of El Toro (Fig. 4.3b), Presta and Caffe (2014) recognized at least four volcanic buildings and a fifth, 6 km to the NE of El Toro (Campo Negro, Fig. 4.3c), constituted them by cinders cones with related lava flows, with the exception of a smaller eruptive center in El Toro that generated only lavic flows.

They identified cones with variable diameters between 600 and 150 m and horseshoe structures for breakage at its base by lava flows, and tabular lavas (andesites and basaltic andesites) lacking preserved textural features of 3–20 m thickness and lengths <100 m to >1 km. Both the scoria cones and the lava fields associated with these eruptive centers respond by their textural, structural, and morphological characteristics to the model of general facies of Strombolian cones (Vespermann and Schmincke 2000).

The development of mafic volcanic fields (e.g., Cerro Morado, Cerro Bitiche, Cerro Negro de Olaroz) must be to the grouping of several nearby centers whose products, mainly effusive, are coalescent or they are juxtaposed. Repeated emissions of lavas caused the stacking of castings emitted from one or several eruptive centers,

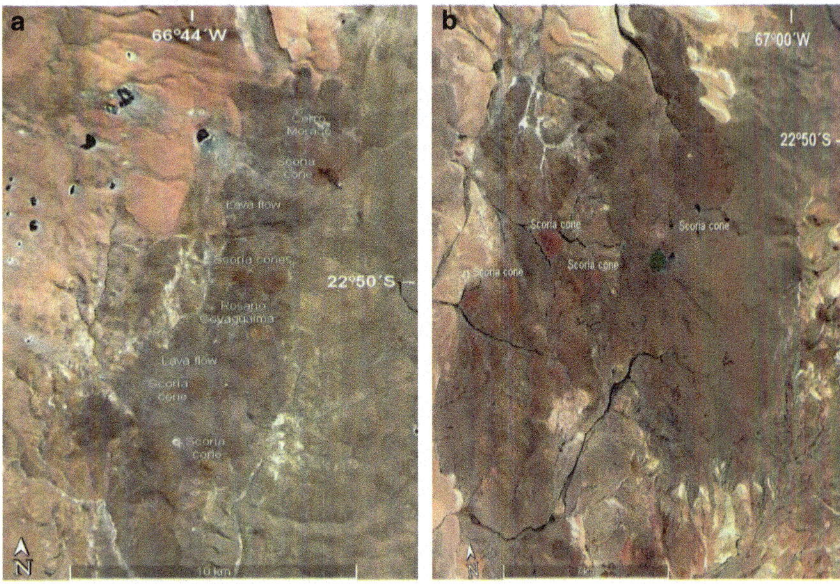

Fig. 4.4 **a** Cerro Morado mafic volcanic field covers a surface of ~100 km^2. It is composed of andesite to basaltic andesite lavas that make up a series of plateaus and scattered scoria cones disposed over those surfaces. **b** Cerro Bitiche mafic field covers an area of ~200 km^2 and comprises scoria cones and extended lava flows. The superposition of lava flows and the intense erosion make difficult the identification of individual lava units

which were filling the space and amalgamating in it. The Cerro Bitiche together with the Cerro Morado represent the largest mafic volcanic fields in the Northern Puna.

The 6.7 Ma "Cerro Morado Andesites" (Coira et al. 1996), studied in detail by Cabrera and Caffe (2009), constitutes a mafic volcanic field (~100 km^2) composed of andesite to basaltic–andesite rocks. This field is integrated into three different sectors (Fig. 4.4a). In each sector, two main geomorphologic features can be recognized: an elevated plateau dominated by lobate lava flows, and a series of scattered scoria cones apparently disposed over the plateau. The elongated shape of the volcanic field, the alignment of the vents (e.g., scoria cones), and the evidence of repeated activity, all suggest that magmas erupted through fissures parallel to main regional lineaments, probably at places where N–S faults met transverse (NW, NE, and _E–W) structures.

Another example of mafic volcanic fields of Northern Puna is the Cerro Bitiche field. It comprises seven scoria cones and extended andesitic lava flows that cover an area of 200 km^2 (Fig. 4.4b). The scoria cones are mainly formed by scoriaceous lapilli of irregular shapes and commonly contain bombs with spindle, ribbon, and subangular forms. They show different degrees of erosion and reach a maximum height of 120 m (Maro and Caffe 2016). The superposition of lava flows and the intense erosion make difficult the identification of individual lava units.

The mafic rocks erupted from the different mafic centers of Northern Puna present aphyric to microporphyric textures. Olivine and subordinated clinopyroxene form the

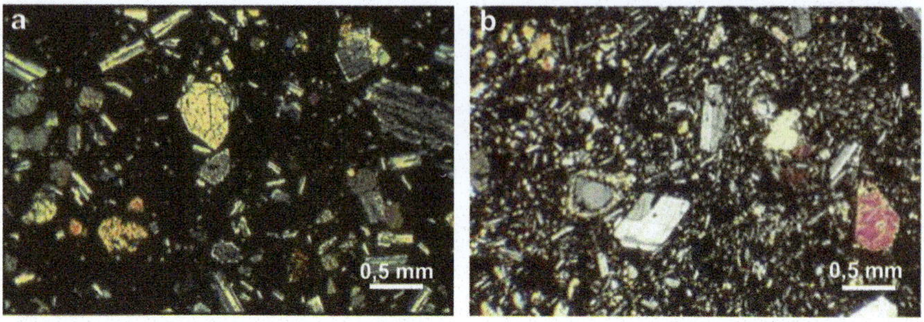

Fig. 4.5 **a** Basaltic andesite with microporphyric texture. Olivine and subordinated clinopyroxene phenocryst in a hyalopilitic groundmass. Crossed polarized light **b** Basaltic andesite with forsterite phenocryst and micro-phenocryst of clinopyroxene, included in a hyalopilitic groundmass together with laths of plagioclase. Xenocrystals of quartz embayed by brown glass and clinopyroxene rims. Crossed polarized light

main mineral assemblage, although rocks with orthopyroxene, plagioclase with or without hydrous minerals (amphibole, biotite), or with only orthopyroxene can also be found (Fig. 4.5a).

The general scarcity of phenocrysts and their sub-skeletal habits support a rapid ascent for most mafic magmas with short or no supracrustal magma chamber histories. Xenocrysts are ubiquitous mainly represented by quartz (Fig. 4.5b).

These mafic rocks consist of basaltic andesites to andesites belonging to the high-K calc-alkaline series. The chemical characteristics of these mafic andesites are close to those of primitive magmas, with moderate to high magnesium numbers (60–67) and Ni and Cr contents (up to 282 ppm and 529 ppm, respectively). An exception is the lava of the eastern center, the Cerro Barro Negro scoria cone, which is a basaltic trachyandesite with a shoshonitic signature (Maro and Caffe 2013). The absence of basaltic compositions is characteristic of Northern Puna mafic volcanism. It is also distinctive of these representatives of Northern Puna volcanism the absence of an evident intraplate type signal (La/Ta < 25, Ba/Ta < 300), as the one that exists in Southern Puna (Kay et al. 1994). In addition, it is observed that mafic lavas of the Northern Puna have higher values of Ba, Rb, Th, Rb/Sr, and Ba/Nb than calc-alkaline rocks of the Southern Puna, as well as they also reach higher values of the relationship $^{87}Sr/^{86}Sr$ (0,705947–0,709613) and $^{143}Nd/^{144}Nd$ (0,512448–0,512279) (Maro et al. 2014).

These contrasts can be explained by differences in the source of the mantle or by the type or amount of cortical contamination between both regions. The second option agrees with a greater compositional contrast observed in the more evolved mafic rocks. Similar discrepancies observed between the ignimbrites of the Northern and Southern Puna were attributed to compositional heterogeneities between the northern and southern basements (Kay et al. 2010). To the contrast in the composition of the crust probably added a greater degree of assimilation of the mafic magmas of Northern Puna as result of the presence of a more extensive partial melting zone in the middle/upper crust, which would explain, on the other hand, the lower volume of mafic rocks in Northern Puna (Maro and Caffe 2016).

4.3 Southern Puna Mafic Volcanism

Late Neogene mafic volcanism in the Southern Puna is concentrated in the region around the Antofalla Salar and Hombre Muerto Salar, Antofalla basin, and extends southward into the Pasto Ventura region (Fig. 4.6). Typical volcanic features are small (<2 km^3) lava flows and cinder cones, some with aprons of ash and lapilli deposits scattered with larger clast, lava domes, and maars grouped in small volcanic fields.

An association of young mafic cones and fissure flows with NNE-trending faults in the Southern Puna has long been recognized (e.g., Hörmann et al. 1973; Coira and Pezzutti 1976; Allmendinger 1986; Coira et al. 1993; Marrett et al. 1994; Marrett and Strecker 2000). Meanwhile, Risse et al. 2013 considered it more likely that magma ascent and emplacement were influenced by the regional NW–SE striking lineament zone and its intersection with NNE–SSW-trending faults bounding the Antofalla Salar basin.

Mafic volcanism in the Southern Puna document a total range of 7.3–0.3 Ma and younger. The onset of back-arc mafic activity is firmly established at least as early as 7 Ma. The data also demonstrate a peak in volcanism at 6–4 Ma, after which activity continued at a lower level, with fluctuations, through the Pleistocene (Risse et al. 2008 and references therein). There is a correspondence in time between the peak of back-arc mafic volcanism and the first major eruptions of large-volume felsic ignimbrites from the nearby Cerro Galán caldera complex. The Galán magmas are largely crustal derived, and it is likely that the increased vigor of back-arc magmatism and the production of extensive crustal melts are both manifestations of enhanced melting in the mantle. The cause of this melting anomaly has been postulated to involve an influx of asthenospheric mantle related to delamination of the lower crust and lithospheric mantle following crustal thickening, or to a steepening/rollback of the subducting plate (e.g., Coira et al. 1993; Kay et al. 1994, 1999).

The Antofagasta de la Sierra basin includes a profuse amount of small mafic centers, those who marginalize the Cerro Galán caldera by the west (Fig. 4.6). They have reached a maximum in the early Pliocene (6–4 Ma) and developed pulses of activity at 3–2 Ma and 1–0 Ma. Among them are recognized centers composed mainly by lava flows and related scoria cones, some lava domes, and minor maars. Examples of them are the young Alumbrera and Laguna centers studied in detail by Báez et al. (2016).

The lava flows associated with the Alumbrera scoria cone (Fig. 4.7a) cover an area of ~42 km^2, have a bulk volume of ~0.3 km^3 and expand southwards along the regional slope, meanwhile the cone, 1.3 km in diameter, rises 164 m above the floor of the basin and has an elliptical morphology in plan view with the maximum elongation oriented NW (Báez et al. 2016).

Fig. 4.6 Distribution of Late Miocene to Pleistocene mafic volcanic centers of the Southern Puna: (**1**) Pasto Ventura, (**2**) Carachipampa, (**3**) Jote, (**4**) Alumbrera, (**5**) Laguna, (**6**) La Falda, (**7**) Nacimientos, (**8**) Chato, (**9**) Juncal Grande, (**10**) Juncalito, (**11**) Diablo, (**12**) Vega Los Colorados, (**13**) Antofalla, (**14**) Chascha, (**15**) F.Catal (S H Muerto), (**16**) Sur Antofalla Salar.

Among the lava flows it is possible to recognize an old group (7.3 Ma, Kay and Coira, unpublished) which present high to moderate levels of wind erosion and the youngest lava flows that are incipient to uneroded and show lava lobes with toothpaste tongues (Fig. 4.7b). The La Laguna cone rises 153 m above the floor of the basin and has an elliptical morphology with the maximum elongation oriented NW and a

Fig. 4.7 **a** Alumbrera old lava flows (7.34 Ma) covered by youngest lava flows expanding along regional slope and associated with small scoria cone. **b** Alumbrera young lava flow showing toothpaste morphology and texture. **c** The Laguna scoria cone on top of "aa" lava flow field. **d** The Laguna volcano, located north of the Alumbrera, shows lava flows (0.3 Ma) with rough surfaces and morphologies such as channels limited by levees and pressure ridges surrounding a small cinder cone (~1 km in diameter)

maximum cone diameter of 1.1 km and has a volume of 0.12 km^3 and an asymmetrical profile (Fig. 4.7c). The lava flows associated with the La Laguna cone occupy an area of 6.8 km^2 and have a bulk volume of 0.12 km^3 (Báez et al. 2016). It is possible to identify an old group of lava flows, highly eroded and a young group, moderately eroded (Fig. 4.7d). The crater of the cone is largely filled by a partly welded spatter and minor fluid bomb deposit that shows a crude bedding dipping inwards toward the crater. To the north and west of the cone, an "aa" lava flow field is conspicuous.

These lavas show rough surfaces and morphologies such as channels limited by levees and pressure ridges. Báez et al. (2016) interpreted that the youngest activity of both volcanoes was characterized by long-lasting eruptions with shifts in the eruptive dynamics in response to a decrease in the mass eruption rate from high-explosive, violent Strombolian to Hawaiian/Strombolian phases. The explosive activity was accompanied by the emission of lava flows from lateral vents, which occurred mostly during the Hawaiian/Strombolian phases. During the waning stage of the eruptions, intermittent phreatomagmatic activity was developed as a result of the sudden access of groundwater into the volcanic conduit.

The emission of successive magma batches with geochemical changes over time in a single scoria cone suggests that both volcanoes are also polymagmatic (sensu Brenna et al. 2010). Temporal hiatuses into the sequences of both centers are identified by significant variations in the degree of erosion and the gap in ages characteristics that correspond to the "monogenetic polycyclic volcanoes" (Németh and Kereszturi 2015).

Other examples of isolated centers in which a succession of lava has been erupted from mouths usually located at the base of scoria cones are the Carachipampa center with young lava flows (0.75 Ma; Kay and Coira unpublished) as large as 8 km with moderate to low thickness (2–6 m) (Fig. 4.8a, c), the older Juncalito (5.03 Ma, Kay and Coira unpublished) with stacking of lavas up to 5 km in length (Fig. 4.8d) and the Diablo which record two groups of lava flow, an old one (4.3 Ma) and a young one uneroded and associated with a cinder cone of 1 km in diameter (Fig. 4.8b).

As clustered centers are recognized: (**a**) the Nacimientos, result of a set of mouths of emission of lava flows of 2.5–1.5 Ma and small scoria cones (Figs. 4.9a, 4.10a) to which they are associated covering an area of 50 km^2; (**b**) Jote, represented by a set of small scoria cones of 0.5–2 km of diameter to which lava flows of 1.5–3 km in length are associated mainly with SSW orientation according to the regional slope (Fig. 4.9b); (**c**) Vega Los Colorados, the older one (4.9 Ma, Marrett et al. 1994) which covers an area of 50 km^2 and gathers a greater number of scoria cones to which stacked lavas are associated, all with evidence of erosion (Fig. 4.10b).

Some clusters have a low density of small and simple scoria cones, lava flows, tuff rings, and maars, aligned along regional NE–SW-trending faults (e.g., the Pasto Ventura; Figs. Fig. 4.11a, 4.12a, b and La Falda; Fig. 4.11b) They are related to a discrete small magma batch, which propagates as a dyke with low magma pressure and spatial and temporal link to traces of preexisting faults (Zhou et al. 2013; Filipovich et al. 2014).

In addition to the tectonic control exerted on the monogenetic centers of the Puna, there is a strong control of the slope of the substrate at the emplacement of the lava flows e.g., Cerro Negro de Purulla, lavaflows of SSW Antofalla Salar (Figs. 4.13, 4.14a, b).

Lava domes are also recognized among the mafic monogenetic centers in Southern Puna e.g., at the foot of Antofalla (Fig. 4.15a), Chascha (Fig. 4.15b), Caipe (Fig. 4.15c), Salar Hombre Muerto (Fig. 4.15d).

They occur in isolation that have elliptical, circular, irregular, and lingual plant view, with diameters of 1–3 km and use to have steep-sided mound and development of transverse surface ridges. They are made of more viscous lavas, mainly of andesitic to basalt andesitic composition, that tend to pile up over their events.

Fig. 4.8 **a** Carachipampa center with young lava flows (0.75 Ma), as large as 8 km covering a surface of ~70 km^2 with moderate to low thickness. Cinder cone with SE flank partially collapsed. **b** Diablo monogenetic polycyclic volcano which recorded an old event represented by moderately eroded lava flows (4.3 Ma) covered by young ones with associated cinder cones with sub-elliptycal plant, WNW oriented as part of the lava flows, suggesting coincidence with the orientation of magma feeding structures. **c** Carachipampa volcano profile (cinder cone and extended lava flows) in the middle of the Antofagasta depression. **d** Juncalito lava flows (5.03 Ma), moderately eroded, with channels limited by levees, emitted from a mouth located at the eastern end and extended in WNW–ESE direction toward the Antofalla Salar

Shoshonitic–andesitic monogenetic volcanoes as San Jerónimo and Negro de Chorrillos are other clear examples of structural control over such monogenetic volcanos. They are localized in an important tectonic transfer zone in the Southern Central Andes (see Fig. 4.16a–d), the NW–SE striking Calama–Olacapato–Toro (COT). NW–SE striking strike-slip faults and NE–SW-trending normal faults accommodated differential shortening between major N–S striking thrust faults on the Puna Plateau along that zone. Petrinovic et al. (2006) present evidences that those centers formed at strike-slip dominated faults. They are about 300–450 m high and 750–950 m in diameter (Petrinovic 1999).

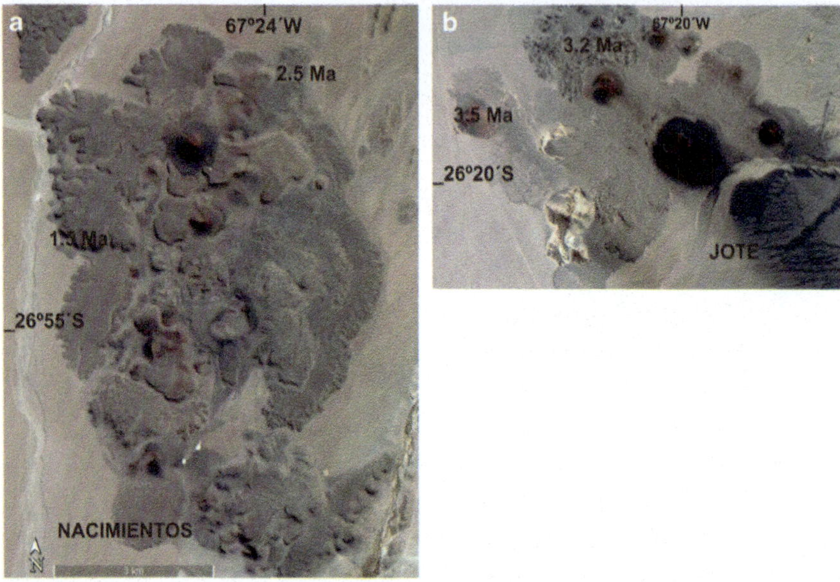

Fig. 4.9 **a** The Nacimientos cluster of lava flows (2.5–1.5 Ma) and small scoria cones covering an area of 50 km2. **b** The Jote cluster of small cones (0.5–2 km of diameter) and "aa" lava flows (3.2–3.5 Ma)

Fig. 4.10 **a** Northern scoria cone of the Nacimientos cluster, with asymmetrical profile, like other of those cones, which are possibly associated with the direction of the prevailing winds during the eruption. **b** The Vega Los Colorados is composed of a cluster of scoria cones and a set the moderately eroded lavas (4.9 Ma) covering an area of 50 km^2

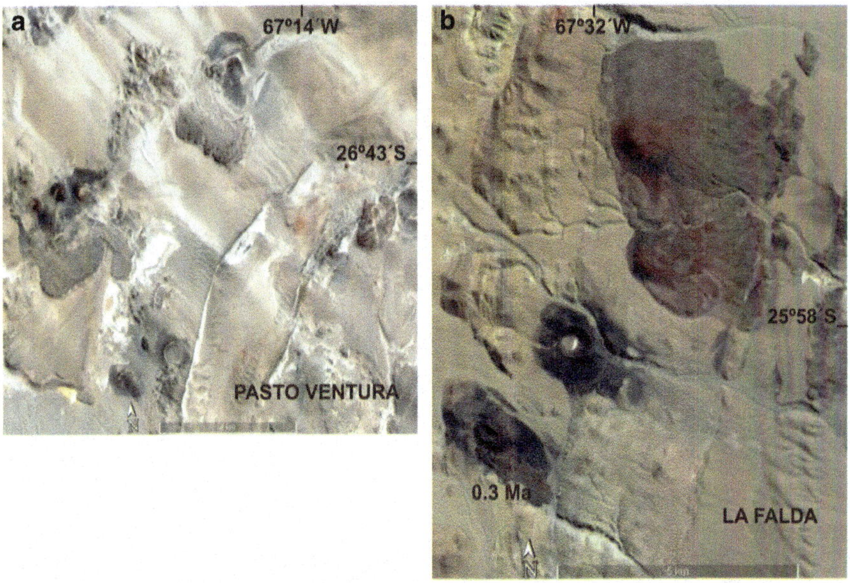

Fig. 4.11 **a** The Pasto Ventura is a cluster of small centers aligned along regional NE–SW-trending faults. It is represented by scoria cones, lava flows, tuff rings, and maars. **b** La Falda consists of small cinder cones and reduced lava flows spatial and temporal associated with pre-existing faults

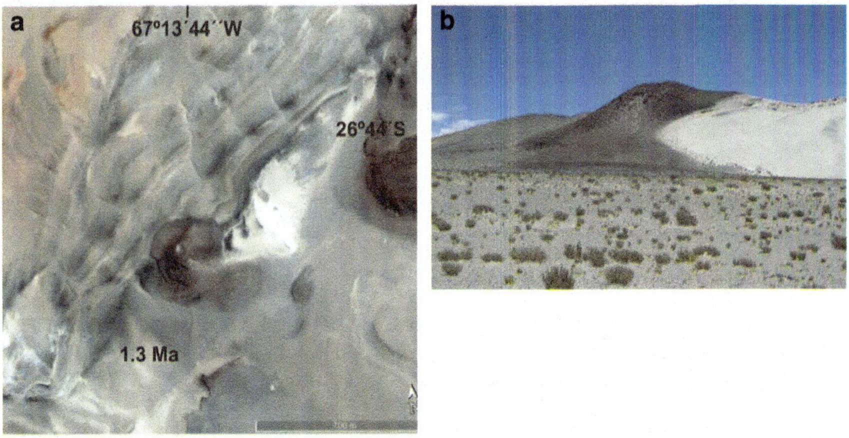

Fig. 4.12 **a** Pasto Ventura very small center (<0.05 km^3, scoria cone and very small lava flows) controlled by NE–SW faulting. **b** Detail of the small center showed in satellital imagen (Fig. 4.12a)

Fig. 4.13 Lava flows of Cerro Negro de Purulla strongly controlled by the inclination of the substrate

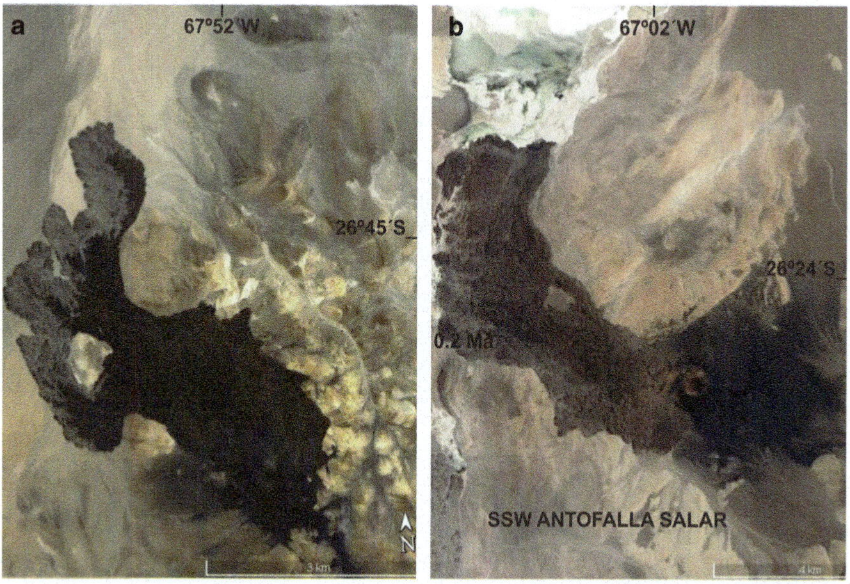

Fig. 4.14 **a** Satellital imagen of Cerro Negro Purulla lava flows controlled by the slope of the hill flank and regional topography. **b** Regional slope and punctual high topographical control on the lava flows of monogenetic center of SSW Salar de Antofalla

The volcanoes formed by accumulations of lavas, bombs, scoria, and ash-fall deposits, whereby lava flows are characterized by block lavas and intermediate "aa" type lavas (Déruelle 1991). The San Jerónimo lavas are found at a distance of 2.5–3 km from the vent controlled by the topography. (Fig. 4.16b), the Negro de Chorrillos lavas occur within a distance of 4 km from the vent (Fig. 4.16c, d).

The mafic lavas of Southern Puna are typically aphyric or weakly porphyritic with a groundmass containing from <3 to ~40 vol. % devitrified glass, which is vesicular in many of the samples. Olivine and clinopyroxene are the main phenocryst

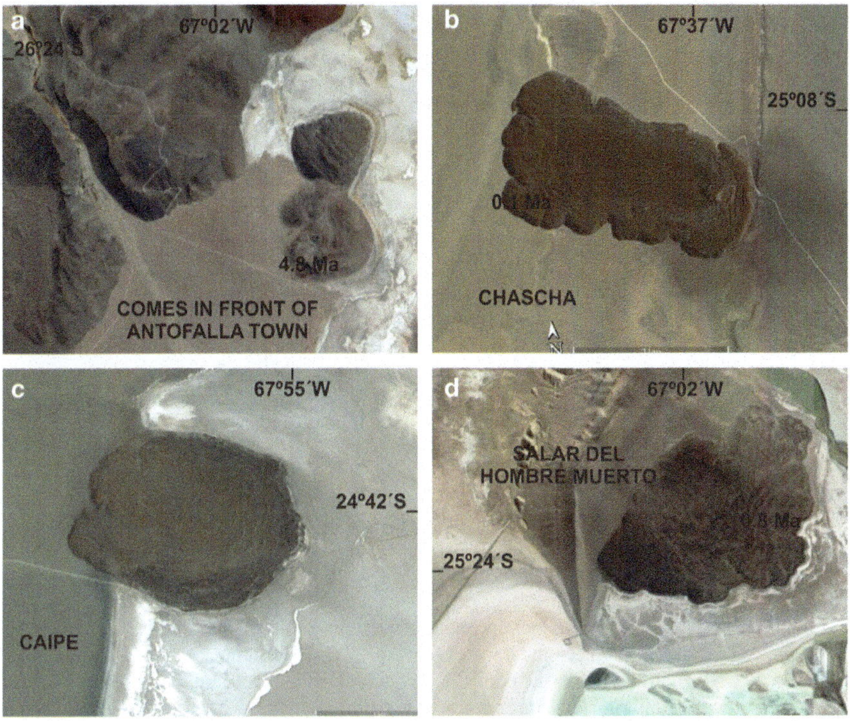

Fig. 4.15 **a** Small lava domes (~1 km in diameter) isolated in the Salar de Antofalla in front of Antofalla town. **b** Vega Chascha lava dome elongated in WNW direction shows lingual plant view with developed transverse surface ridges and an emission mouth located in the east end. **c** Caipe lava dome, marginal to the Arizaro Salar, is subcircular with a diameter of 1.6 km, steep-sided mound, and the emission mouth in the west end. **d** Lava dome (0.8 Ma) located in the middle of Hombre Muerto Salar. Its emission has been controlled by NNE regional fault that tilted the Miocene sedimentary sequences

phases. Olivine is more abundant than pyroxene in all, but not in the most silicic samples. Plagioclase is exclusively a groundmass phase in the mafic lavas, and it commonly defines a flow lineation (Fig. 4.17a). Only the most evolved, andesitic rocks (e.g., with 60% SiO_2) contain clinopyroxene and plagioclase phenocrysts, as well as minor amounts of orthopyroxene and amphibole (Fig. 4.17b). The most abundant groundmass minerals are clinopyroxene, plagioclase, Fe Ti oxides, and, locally, olivine in the least-evolved rocks. Accessory minerals, which mostly occur as inclusions in these phenocrysts, include chromian spinel, pentlandite, pyrrhotite, and apatite.

The main compositional features of the Southern Puna mafic volcanic rocks have been described in previous studies (Francis et al. 1989; Knox et al. 1989; Kay et al. 1994, 1999; Kraemer et al. 1999; Drew et al. 2009). The most common rocks in the back-arc mafic suite of the Puna are basaltic andesites and basaltic trachyandesites, with minor amounts of basalt, trachybasalt, and andesites. Knox et al. (1989),

Fig. 4.16 **a** Shoshonitic–andesitic monogenetic volcanoes as San Jerónimo and Negro de Chorrillos are localized in the important tectonic transfer zone, the NW–SE striking Calama–Olacapato–Toro (COT). **b** The San Jerónimo volcanic edifice ~300 m high is emplaced on top of Miocene Aguas Calientes Caldera ignimbrites. Its cone is formed by accumulations of bombs, scoria, and ash-fall deposits. The lava flows were mainly extruded from the base of the cone and are found at a distance of 8–10 km from the vent toward the SW. **c** The Negro de Chorrillos center (0.45 Ma) located on top of Ordovician basement is spatially associated with NW–SE vertical strike-slip faults. Its blocky to "aa" lava flowed following the valley along 4 km in the direction of San Antonio de los Cobres. **d** The Negro de Chorrillos volcanic edifice showing debris avalanches affecting the cone

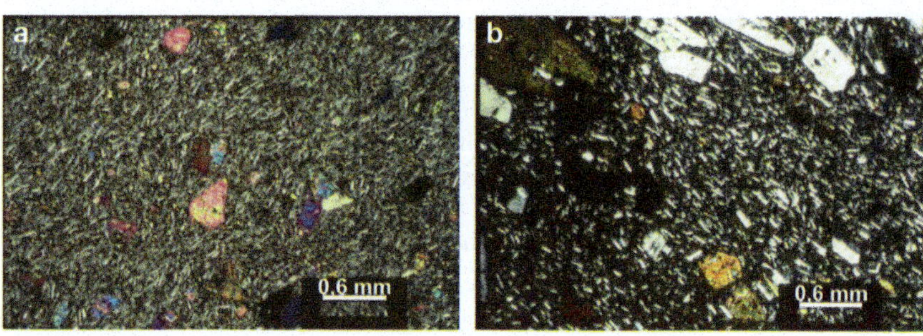

Fig. 4.17 **a** Basaltic andesite weakly porphyritic with olivine and clinopyroxene as the main phenocryst phases. Plagioclase is exclusively a groundmass phase associated with clinopyroxene, Fe Ti oxides, and locally olivine. **b** Andesite containing clinopyroxene and plagioclase phenocrysts, as well as minor amounts of orthopyroxene and amphibole in a groundmass of desvitrified glass microlites of plagioclase and minor clinopyroxene and Fe Ti oxides. (Crossed polarized light)

Kay et al. (1994) distinguished three compositional series based on incompatible trace element ratios (REE, La/Ta, Ba/Ta) and K2O contents. The most common and widespread is the Calc-Alkaline (CA) series, distinguished by relatively high ratios of La/Ta (>25) and Ba/Ta (>300), which is interpreted as having a strong influence of fluid-transported components from the subducted slab. The within-plate series has relatively low values of the La/Ta and Ba/Ta ratios, suggesting minor influence of slab-derived components in the source. The within-plate series is restricted geographically to the central part of the back-arc region, west of Cerro Galan (Kay et al. 1994). The third, shoshonitic series is distinguished by high K2O contents. It occurs farther north in the back-arc region, between 24 and 25°S latitude (e.g., San Jerónimo and Negro de Chorrillos). The calc-alkaline and within-plate groups are about equally represented in the Southern Puna. They are in the range of about 50–57 wt % SiO_2, about 5–10 wt % MgO, 57–70 Mg, Ni (150–210 ppm), and Cr (400–550 ppm), which are close to acceptable values for primary magmas. The trace elements distinction between the calc-alkaline and within-plate series shows up in deeper negative Nb-Ta troughs at equivalent MgO for the former in the mantle-normalized trace element patterns. This difference is independent of the degree of differentiation (e.g., MgO content) and agrees with the Ba/Ta and La/Ta ratio distinction (Risse et al. 2013).

References

Allmendinger RW (1986) Tectonic development, southeastern border of the Puna Plateau, northwestern Argentine Andes. Geol Soc Am Bull 97:1070–1082

Báez W, Carrasco Nuñez G, Giordano G, Viramonte JG, Chiodi A (2016) The polycyclic scoria cones of the Antofagasta de la Sierra basin, Southern Puna Plateau, Argentina. In: Németh K, Carrasco- Núñez G, Aranda-Gómez JJ, Smith IEM (eds) Monogenetic volcanism. Geological Society of London, Special Publications 446, pp 311–336

Brenna M, Cronin SJ, Smith IE, Sohn YK, Németh K (2010) Mechanisms driving polymagmatic activity at a monogenetic volcano, Udo, Jeju Island, South Korea. Contrib Miner Petrol 160:931–950

Cabrera AP, Caffe PJ (2009) The Cerro Morado Andesites: volcanic history and eruptive styles of a mafic volcanic field from northern Puna, Argentina. J S Am Earth Sci 28:113–131

Coira B, Pezzutti N (1976) Vulcanismo cenozoico en el ámbito de Puna Catamarqueña. Rev Asociación Geológica Argentina XXXI(1):33–52

Coira B, Kay SM, Viramonte JG (1993) Upper Cenozoic magmatic evolution of the Argentine Puna – a model for changing subduction geometry. Int Geol Rev 35:677–720

Coira B, Caffe P, Mahlburg Kay S, Diaz A, Ramírez A (1996) Complejo volcánico de Vilama – Sistema caldérico del Cenozoico superior en Puna, Jujuy. In: XIII Congreso Geológico Argentino, Actas 3, Buenos Aires, pp 603–620

Déruelle B (1991) Petrology of quaternary shoshonitic lavas of northwestern Argentina. Geol Soc Am Spec Pap 265:201–216

Drew ST, Ducea MN, Schoenbohm LM (2009) Mafic volcanism on the Puna plateau, NW Argentina: Implications for the lithospheric composition and evolution with an emphasis on lithospheric foundering. Lithosphere 1:305–318

Filipovich R, Santillán A, Báez W, Viramonte JG (2014) Caracterización del volcanismo monogénico básico de la región de Pasto Ventura, Puna Austral. In: XIX Congreso Geológico Argentino, Actas: S24-1-6, Córdoba

Francis PW, Sparks RSJ, Hawkesworth CJ, Thorpe RS, Pyle DM, Tait SR, Mantovani MS, McDermott F (1989) Petrology and geochemistry of volcanic rocks of the Cerro Galan caldera, northwest Argentina. Geol Mag 126:515–547

Guzmán S, Petrinovic IA, Brod JA (2006) Pleistocene mafic volcanoes in the Puna-Cordillera Oriental boundary, NW-Argentina. J Volcanol Geotherm Res 158:51–69

Hooper DM, Sheridan MF (1998) Computer simulation models of scoria cone degradation. J Volcanol Geotherm Res 83:241–267

Hörmann KP, Pichler KH, Zeil W (1973) New data on the young volcanism in the Puna of NW-Argentina. Geologische Runschau 62:397–428

Kay SM, Coira B, Viramonte JG (1994) Young mafic backarc volcanic rocks as indicators of continental lithospheric delamination beneath the Argentine Puna Plateau, Central Andes. J Geophys Res 99:24323–24339

Kay SM, Mpodozis C, Coira B (1999) Neogene magmatism, tectonism, and mineral deposits of the Central Andes (22–33_S). In: Skinner B (ed) Geology and ore deposits of the Central Andes, Society of Economic Geology Special Publication 7, pp 27–59

Kay S, Coira B, Caffe P, Chen C (2010) Regional chemical diversity, crustal and mantle sources and evolution of Central Andean Puna plateau ignimbrites. J Volcanol Geotherm Res 198:81–110

Kereszturi G, Németh K (2012) Monogenetic basaltic volcanoes: genetic classification, growth, geomorphology and degradation. In: Németh K (ed) Updates in volcanology – new advances in understanding volcanic systems. InTech, pp 3–88

Knox WJ Jr, Kay SM, Coira B (1989) Geochemical evidence on the origin of Quaternary basaltic andesites of the Puna, northwestern Argentina. Rev de la Asociación Geológica Argentina 44:194–206

Kraemer B, Adelmann D, Alten M, Schnurr W, Erpenstein K, Kiefer E, van den Bogaard P, Görler K (1999) Incorporation of the Paleogene foreland into the Neogene Puna Plateau; the Salar de Antofalla area, NW Argentina. J S Am Earth Sci 12:157–182

Marrett R, Emermann SH (1992) The relations between faulting and mafic magmatism in the Altiplano-Puna plateau (central Andes). Earth Planet Sci Lett 112:53–59

Marrett RA, Strecker MR (2000) Response of intracontinental deformation in the Central Andes to late Cenozoic reorganization of South American plate motions. Tectonics 19:452–467

Marrett RA, Allmendinger RW, Alonso RN, Drake RE (1994) Late Cenozoic tectonic evolution of the Puna Plateau and adjacent foreland, northwestern Argentine Andes. J S Am Earth Sci 7:179–207

Maro G, Caffe PJ (2012) Volcanismo máfico terciario de la Puna jujeña, los Cerros Negros de Jama. In: López JP, Grosse P, Bellos LI, Baéz MA, Fogliata AS, Sardi FG (eds) Aportes al Magmatismo y Metalogénesis asociada de la República Argentina I. Serie Correlación Geológica, Tucumán, Argentina - INSUGEO, vol 28, pp 51–72

Maro G, Caffe PJ (2013) Primer registro de lavas shoshoníticas en Puna norte, el cerro Barro Negro. Actas 2do. Simposio sobre Petrología Ignea y Metalogénesis Asociada, San Luis, p 51

Maro G, Caffe PJ, Trumbull R, Romer R (2014) Andesitas magnesianas del volcanismo monogenético neógeno de la Puna jujeña, Andes Centrales. XIX Congreso Geológico Argentino, Actas: S24-1-2, Córdoba

Maro G, Caffe PJ, Jofré C (2016) Xenolitos ultramáficos en lavas máficas neógenas de la Puna Norte. Rev de la Asociación Geológica Argentina 73:282–293

Maro G, Caffe PJ, Báez W (2017) Volcanismo monogenético máfico cenozoico de la Puna. In: Muruaga CM, Grosse P (eds) Ciencias de la Tierra y Recursos Naturales del NOA. Relatorio del XX Congreso Geológico Argentino, San Miguel de Tucumán, pp 548–577. ISBN 978-987-42-6666-8

Montero López MC (2009) Estructura y magmatismo Neógeno-Cuaternarios en la Sierra de San Buenaventura (Catamarca): su vinculación con la terminación austral de la Puna. Ph.D. Thesis, Universidad Nacional de Salta, Salta, Argentina, p 255

Németh K, Kereszturi G (2015) Monogenetic volcanism: personal views and discussion. Int J Earth Sci 104:2131–2146

Norini G, Báez W, Becchio R, Viramonte J, Giordano G, Arnosio M, Piton A, Groppelli G (2013) The Calama–Olacapato–El Toro fault system in the Puna Plateau, Central Andes: geodynamic implications and stratovolcanoes emplacement. Tectonophysics 608:1280–1297

Petrinovic IA (1999) La Caldera de colapso del Cerro Aguas Calientes, Salta; República Argentina; Evolución y Esquema Estructural. In: Colombo F, Queralt I, Petrinovic IA (eds) Geología de los Andes Centrales Meridionales: El Noroeste Argentino. Acta Geol Hisp 34:243–255

Petrinovic IA, Riller U, Brod JA, Alvarado G, Arnosio M (2006) Bimodal volcanism in a tectonic transfer zone: evidence for tectonically controlled magmatism in the southern Central Andes, NW Argentina. J Volcanol Geotherm Res 152:240–252

Presta JF, Caffe PJ (2014) Historia eruptiva de los volcanes monogenéticos de El Toro (23° 05′S – 66° 42′W), Puna norte, Argentina. Andean Geol 41:142–173

Riller U, Oncken O (2003) Growth of the central Andean plateau by tectonic segmentation is controlled by the gradient in crustal shortening. J Geol 111:367–384

Risse A, Trumbull RB, Coira B, Kay SM, van den Bogaard P (2008) 40Ar/39Ar geochronology of basaltic volcanism in the back-arc region of the southern Puna plateau, Argentina. J S Am Earth Sci 26:1–15

Risse A, Trumbull RB, Kay SM, Coira B, Romer RL (2013) Multi-stage evolution of late Neogene mantle-derived magmas from the central Andes back-arc in the Southern Puna Plateau of Argentina. J Petrol 54(10):1963–1995

Schwab K, Lippolt H (1974) K-Ar mineral ages and late Cenozoic history of the Salar de Cauchari area (Argentina Puna). In: González F (ed) Proceedings of the symposium on Andean and Antarctic Volcanology problems, Santiago. International Association of Volcanology and Chemistry of the Earth's Interior, Rome, pp 697–714

Vespermann D, Schmincke HU (2000) Scoria cones and tuff rings. In: Sigurdsson H, Houghton B, McNutt SR, Rymer H, Stix J (eds) Encyclopedia of Volcanoes. Academic Press, San Diego, pp 683–694

Part II
Paleozoic Magmatism

Volcanic complexes showing well exposed and preserved outcrops along the Puna Argentina and to the south, at the north of the Famatina System, in the context of the Famatinian Orogen, are analyzed at the next chapters. Lower Paleozoic units take part of the proto-Andean margin, characterized by an Early Ordovician magmatism extending from Venezuela to northeast Patagonia as was indicated by Pankhurst et al. (2006), Chew et al. (2007), Alasino et al. (2016), Van der Lelij et al. (2016), García-Ramírez et al. (2017), Ramos (2018), among others. This magmatic–tectonic arc was constructed during the Famatinian Orogenic Cycle (Aceñolaza & Toselli 1976).

The Famatinian magmatic arc along the proto-margin of western Gondwana has complex and variable tectonic settings at its different segments. They form an almost semicontinuous belt of arc related rocks with ages varying from 490 to 460 Ma, which correlates with the time span of the Famatinian Orogeny (Ramos 2018). Famatinian plutonic and volcanic rocks along a 500 km wide transect between 27° and 33°S, including the Sierra de Famatina to the north (Rapela et al. 2018). At 28°S have well exposed units prooving changes in paleo-depth and showing the deepest plutonic levels of the arc. In contrast, volcanic complexes and coeval ensialic basins dominate the Puna plateau to the north (Otamendi et al. 2012; Rapela et al. 2018, and references therein).

Along the north and central Andes, Famatinian Belt outcrops are narrow and essentially subparallel to the modern Andean chain, although some of them are largely covered of younger sedimentary units or are only represented by small remnants. Nevertheless, in Puna, the Famatinian records were integrated in the Western and Eastern Magmatic belts composed by Ordovician volcanic–sedimentary deposits and granitoids emplaced mainly to the south. In Sierras Pampeanas, well-exposed N-S striking ranges of hundreds of kilometres are showing a great variety of Famatinian units.

The tectonic setting of the Famatinian Belt during Early Palaeozoic has been a matter of debate in the last decades, however paleontological, petrological, geochronological and paleomagnetic data on the Famatinian Orogeny records support subduction of oceanic lithosphere along the western margin of Gondwana (e.g.,

Rapela et al. 1992; Toselli et al. 1996, Benedetto 1998; Rapalini et al. 1999; Astini and Dávila 2004; Ramos 2018, and references therein).

The volcanic island setting for the Famatina Range (located in the central sector of the Famatina Belt) was initially suggested by Aceñolaza and Tosselli (1984). Subsequently, some authors have favored a back-arc model on petrological basis (Mannheim 1993; Clemens 1993 and Toselli et al. 1996). Others, analyzing the stratigraphic and sedimentologic records have indicated an intra-arc basin evolution (Mangano and Buatois 1996; Cisterna et al. 2010a). Isotopic and geochemical studies on Famatina Early Ordovician granitoids documented the evolution of that magmatic arc on a thickened continental crust (Pankhurst et al. 2000), instead of a genesis under an island arc regime with subduction of oceanic crust under another oceanic crust. The extension of this magmatic arc northward in the Puna magmatic belts have been postulated by Coira (1975, 1979), Coira et al. (1982), Ramos (1988), Conti et al. (1996), and Cisterna et al. (2010b, 2017). Taking into account the geochemical study of the basic volcanism developed along the Las Planchadas and Narváez hills, Cisterna et al. (2017) postulated for the evolution of the volcanism transition from conditions of island arc to back-arc along an overriding plate progressively thicker and comparable to some segments of the actual Central America Volcanic Arc.

Geochemical characteristics of the volcanism in the Western and Eastern Magmatic belts forming the Puna region, indicate back-arc basins operating in the Tremadocian early stages. During Early-Middle Arenigian, the volcanism in the Western Magmatic Belt shows geochemical characteristics of arc to back-arc developed on a thin continental crust and in the Eastern Magmatic Belt, basic rocks with intraplate signature and siliceous rocks with cortical filiation indicate a tectonic environment where the alkaline magmatism is the result of decompression of the mantle along pull-apart regions, while important cortical fusion takes place due to the thermal anomaly related to thinning (Coira 2008; Coira et al. 2009).

In the northern portion of the Famatina System and southern Puna, the Ordovician granitoids and volcanic successions are usually covered by thick Late Paleozoic siliciclastic units (e.g., Turner 1967). Recent studies carried out in this region indicate that rhyolitic rocks, previously mapped as Ordovician and included in the Famatinian magmatic arc (Turner 1967; Mpodozis et al. 1997; Rubiolo et al. 2001; Coira et al. 2005), have Carboniferous age, pointing out to a volcanic event in the region (e.g., Martina et al. 2011). The available Mississippian ages (348-342 Ma) obtained by the latter authors in the Carboniferous volcanic units outcropping at Cazadero Grande (Catamarca) and Cerro Pabellones (La Rioja) allowed to consider these units associated to crustal extension in a retro-arc environment; they also being related to coeval A-type granites (340-353 Ma) exposed to the east, in Sierras Pampeanas (Grosse et al. 2009; Dahlquist et al. 2006, 2013). The latter authors have reported isotope data of zircons from Lower Carboniferous granites that indicate their derivation from recycling of juvenile crustal material, in accordance with regional studies of Willner et al. (2008).

Along the Chaschuil valley were reported U-Pb SHRIMP ages of 348±3–342±1 Ma (Martina et al. 2011) and 320±2 Ma (Cisterna et al. 2013; Zappettini et al. 2015) by the Carboniferous volcanic records. They represent bimodal volcanism with basalts displaying features comparable to transitional MORB and within-plate tholeiites, with contributions from a mantle source affected by crustal contamination (Coira et al. 2016) and the acid members showing geochemical affinities to within-plate magmas and composition suggesting derivation from continental crustal material with mantle source interaction or a juvenile essentially mantle derived crust (Coira et al. 2016).

References

Aceñolaza FG, Toselli AJ (1976) Consideraciones estratigráficas y tectónicas sobre el Paleozoico inferior del Noroeste Argentino. II Congreso Latinoamericano de Geología, Caracas, Actas 2:755–764

Aceñolaza FG, Toselli A (1984) Lower Ordovician volcanismin North West Argentina. In: Bruton DI (ed) Aspects of the Ordovician System. Palaeontological Contributions from the University of Oslo, Oslo 295:203–209

Alasino PH, Casquet C, Pankhurst RJ, Rapela CW, Dahlquist JA, Galindo C, Larrovere MA, Recio C, Paterson SR, Colombo F, Baldo EG (2016) Mafic rocks of the Ordovician Famatinian magmatic arc (NW Argentina): new insights into the mantle contribution. Geol Soc Am Bull 128:1105–1120

Astini RA, Dávila FM (2004) Ordovician back arc foreland and Ocloyic thrust belt development on the western Gondwana margin as a response to Precordillera terrane accretion. Tectonics 23, TC4008, https://doi.org/10.1029/2003TC001620

Benedetto JL (1998) Early Paleozoic brachiopods and associated shelly faunas from western Gondwana: its bearing on the geodynamic history of the pre- Andean margin. In: Pankhurst R, Rapela CW (eds) Theproto - Andean margin of Gondwana. Geol Soc Spec Pub 142:57–83

Cisterna CE, Coira B, Décima F (2010a) Efusiones subácueas del arco volcánico ordovícico en el norte del Sistema de Famatina. Rev Asoc Geol Argentina 66:223–235

Cisterna CE, Coira B, Koukharsky M (2010b) Sucesiones volcánicas-sedimentarias tremadocianas y arenigianas en la sierra de Las Planchadas-Narváez: registros evolutivos del arco magmático famatiniano. Rev Asoc Geol Argentina 66:178–191

Cisterna CE, Coira B, Koukharsky M, Zappettini E, Santos JO (2013) Volcanismo carbonífero posorogénico en el río de Las Lozas, valle de Chaschuil, Catamarca, Argentina: nuevas evidencias de extensión en el borde occidental de Gondwana. In: 6 Simposio Argentino del Paleozoico Superior Actas 1, Buenos Aires, pp 8

Cisterna CE, Koukharsky M, Coira B, Günter C, Ulbrich H (2017) Arenigian tholeiitic basalts in the Famatina Ordovician basin, northwestern Argentina: emplacement conditions and their tectonic significance. Andean Geology 44(2):123–146

Chew DM, Schaltegger U, Košler J, Whitehouse MJ, Gutjahr M, Spikings RA, Miškovíc A (2007) U–Pb geochronologic evidence for the evolution of the Gondwanan margin of the north-central Andes. Geol Soc Am Bull 119:697–711

Clemens K (1993) Sedimentología, proveniencia y desarrollo geotectónico del Sistema de Famatina en el noroeste de Argentina durante el Paleozoico inferior. In: Congreso Geológico Argentino, 2, Actas 1, Mendoza, pp 310–321

Coira B (1975) Ciclo efusivo ordovícico registrado en la Formación Acoite, Abra Pampa, Provincia de Jujuy, Argentina. In: 2 Congreso Iberoamericano de Geología Económica, Actas 1, Buenos Aires, pp 37–56

Coira B (1979) Descripción Geológica de la Hoja 3C Abra Pampa, provincia de Jujuy. Servicio Geológico Nacional Boletín 170:1–90

Coira B, Davidson J, Mpodozis C, Ramos V (1982) Tectonic and magmatic evolution of the Andes of northern Argentina and Chile. Earth Sci Rev 18:303–332
Coira B, Cisterna CE, Koukharsky M (2005) Sucesión hiloclástica-lávica, bimodal, ordovícica de la quebrada de las Lozas, Puna Catarmarqueña, Argentina. In: 16 Congreso Geológico Argentino Actas 4, La Plata, pp 679–686
Coira B (2008) Volcanismo del Paleozoico inferior en la Puna Jujeña. In: Coira B, Zappettini EO (eds) Geología y Recursos Naturales de la Provincia de Jujuy, Relatorio del XVII Congreso Geológico Argentino, Asociación Geológica Argentina. Buenos Aires, pp 140–154
Coira B, Koukharsky M, Ribeiro Guevara S, Cisterna CE (2009) Puna (Argentina) and Northern Chile Ordovician basic magmatism: A contribution to the tectonic setting. J S Am Earth Sci 27:24–35
Coira B, Cisterna CE, Ulbrich H, Cordani H (2016) Extensional Carboniferous magmatism at the western margin of Gondwana: Las Lozas valley, Catamarca, Argentina. Andean Geology 43(1):105–126
Conti CM, Rapalini AE, Coira B, Koukharsky M (1996) Paleomagnetic evidence of an early Paleozoic rotated terrane in nothwestArgentina: A clue from Gondwana – Laurentia interactions? Geology 24:953–956
Dahlsquist JA, Pankhurst RJ, Rapela CW, Casquet C, Fanning CM, Alasino P, Baez M (2006) The San Blas Pluton: An example of Carboniferous plutonism in the Sierras Pampeanas, Argentina. J S Am Earth Sci 20:341–350
Dahlquist JA, Pankhurst RJ, Gasching RM, Rapela CW, Casquet C, Alasino PH, Galindo C, Baldo EG (2013) Hf and Nd isotopes in Early Ordovician to Early Carboniferous granites as monitors of crustal growth in the Proto-Andean margin of Gondwana. Gondwana Res 23:1617–1630
García-Ramírez CA, Rey-León V, Valencia V (2017) Ortoneises en la Franja Silos-Babega, Macizo de Santander, Colombia: evidencias de la orogenia famatiniana en los Andes del norte. Andean Geology 44:307–327
Grosse P, Söllner F, Báez M, Toselli A, Rossi J, De La Rosa JD (2009) Lower Carboniferous post-orogenic granites in central-eastern Sierra de Velasco, Sierras Pampeanas, Argentina: U-Pb monazite geochronology, geochemistry and Sr-Nd isotopes. J Earth Sci 98(5):1001–1025
Mángano MG, Buatois LA (1996) Shallow marine event sedimentation in a volcanic arc-related setting: the Ordovician Suri Formation, Famatina Range, northwest Argentina. Sed Geol 105:63–90
Mannheim R (1993) Génesis de las volcanitas eopaleozoicas del Sistema del Famatina, Noroeste de Argentina. In: 12 Congreso Geológico Argentino y 2 Congreso de Exploración de Hidrocarburos Actas 4, Mendoza, pp 147–155
Martina F, Viramonte JM, Astini RA, Pimentel MM, Dantas E (2011) Mississippian volcanism in the south-central Andes: New U-Pb SHRIMP zircón geochronology and whole-rock geochemistry. Gondwana Res 19(2):524–534
Mpodozis C, Kay SM, Gardeweg M, Coira B (1997) Geología de la región de Valle Ancho-Laguna Verde (Catamarca, Argentina): una ventana al basamento del extremo sur de la zona volcánica de los Andes Centrales. In: 7 Congreso Geológico Chileno Actas 3, Antofagasta, pp 1689–1693
Otamendi JE, Ducea MN, Bergantz GW (2012) Geological, petrological and geochemical evidence for progressive construction of an arc crustal section, Sierra de Valle Fértil, Famatinian arc, Argentina. J Petrol 53:761–800
Pankhurst R, Rapela C, Fanning H (2000) Age and origin of coeval TTG, I-and S-Type granites in the Famatinian belt of NW Argentina. Trans R Soc Edinb Earth Sci 91:151–168
Pankhurst RJ, Rapela CW, Fanning CM, Márquez M (2006) Gondwanide continental collision and the origin of Patagonia. Earth Sci Rev 76:235–257
Ramos VA (1988) Late Proterozoic-Early Paleozoic of South America: a collisional history. Episodes 11(3):168–175
Ramos VA (2018) The Famatinian orogen along the protomargin of Western Gondwana: Evidence for a nearly continuous Ordovician magmatic arc between Venezuela and Argentina. In: Folguera A, Contreras Reyes E, Heredia N, Encinas AB, Iannelli S, Oliveros V, Dávila F, Collo G,

Giambiagi L, Maksymowicz A, Iglesia Llanos MP, Turienzo M, Naipauer M, Orts D, Litvak V, Alvarez O, Arriagada C (eds) The evolution of the Chilean-Argentinean Andes. Springer Earth System Sciences 6:154–183

Rapalini AE, Astini RA, Conti CM (1999) Paleomagnetic constrains on the evolution of Paleozoic suspect terranes from South America. In: Ramos VA, Keppie JD (eds) Laurentia e Gondwana Connections before Pangea. Geol Soc Am Spec Pap 336:171–182

Rapela CW, Coira B, Toselli A, Saavedra J (1992) The lower Paleozoic magmatism of southwestern Gondwana and the evolution of Famatinian orogen. International Geology Review 34:1142–1181

Rapela CW, Pankhurst RJ, Casquet C, Dahlquist JA, Fanning CM, Baldo EG, Galindo C, Alasino PH, Ramacciotti CD, Verdecchia SO, Murra JA, Basei MA (2018) A review of the Famatinian Ordovician magmatism in southern South America: Evidence of lithosphere reworking and continental subduction in the early proto-Andean margin of Gondwana. Earth Sci Rev, https://doi.org/10.1016/j.earscirev.2018.10.006

Rubiolo D, Cisterna CE, Villeneuve M (2001) Edad U/Pb del granito de Las Angosturas en la sierra de Narváez (Sistema de Famatina, provincia de Catamarca). In: 15 Congreso Geológico Argentino, Actas 1, Calafate, pp 359–362

Toselli AJ, Durand FR, Rossi de Toselli JN, Saavedra J (1996) Esquema de evolución geotectónica y magmática eopaleozoica del Sistema de Famatina y sectores de Sierras Pampeanas. In: 13 Congreso Geológico Argentino, Actas 5, Buenos Aires, pp 443–462

Turner JM (1967) Descripción Geológica de la Hoja Geológica 13b Chaschuil, provincias de Catamarca y la Rioja. Servicio Geológico Nacional Boletín 106:1–123

Van der Lelij R, Spikings R, Ulianov A, Chiaradia M, Mora A (2016) Palaeozoic to Early Jurassic history of the northwestern corner of Gondwana, and implications for the evolution of the Iapetus, Rheic and Pacific Oceans. Gondwana Res 31:271–294

Willner AP, Gerdes E, Massone H-J (2008) History of crustal growth and recycling at the Pacific convergent margin of South America at latitudes 29°–36° S revealed by a U–Pb and Lu–Hf isotope study of detrital zircon from late Paleozoic accretionary systems. Chem Geol 253:114–129

Zappettini E, Coira B, Santos J, Cisterna CE, Belousova E (2015) Combined U-Pb and Lu-Hf isotopes from the Las Lozas volcanics, Chaschuil Valley, NW Argentina: evidence of lower Pennsylvanian extensional volcanism in western Gondwana. J S Am Earth Sci 59:13–18

Chapter 5
Ordovician Volcanism in the Northern Famatina System

Abstract The Suri–Las Planchadas Volcanic–Sedimentary Complex at the northwestern of the Famatina System, Famatinian Orogen, comprises an Early to Medium Arenigian submarine and subaerial succession with a variety of mafic to felsic volcanic lithofacies. The architecture of these units includes mafic to intermediate submarine massive-hyaloclastic flows, felsic lavas, pyroclastic and pyroclastic rich volcaniclastic deposits, intrusions of dykes, cryptodomes and rhyolitic lavas. The diversity of volcanic lithofacies recognized along the region mainly reflects variations in processes, environment of emplacements, and eruption styles. Basalts are the dominant lavas, massive, fragmented, and interbedded with hyaloclastites and resediment hyaloclastites (basaltic breccias), testifying subaqueous emplacement conditions. Locally, basalts are pillowed and forming pillow-breccias deposits. Interflow siltstones and sandstones, comprising centimetre to decimetre beds, planar to wedge-shaped form, are common among the lavas and breccias. Nearly 2 km of the exposed succession is dominated by subaqueous-resedimented volcaniclastic siltstones, sandstones, and breccias with abundant pyroclastic components. The explosive activity is testified by the occurrence of minor subaerial pyroclastic deposits (lithic tuff), abundant shards, and pumice fragments in resedimented deposits. The subaqueous-resedimented volcaniclastic facies were defined as high concentration volcaniclastic debris flows and high- to low-density turbidity currents deposits.

Keywords Ordovician · Subaqueous · Hyaloclastites · Volcaniclastic breccias

5.1 Introduction

The Ordovician rocks in northern Famatina System (27° and 31° S), Sierras Pampeanas (Fig. 5.1) comprise the most complete and continuous Tremadocian and Arenigian volcanic-sedimentary successions. They show very extended outcrops with no significant structural modifications, being possible to understand their geological relations and to interpret primary structures and textures.

B. L. L. Coira and C. E. Cisterna, *Textures, Structures and Processes of Volcanic Successions*, Springer Earth System Sciences,
https://doi.org/10.1007/978-3-030-52010-6_5

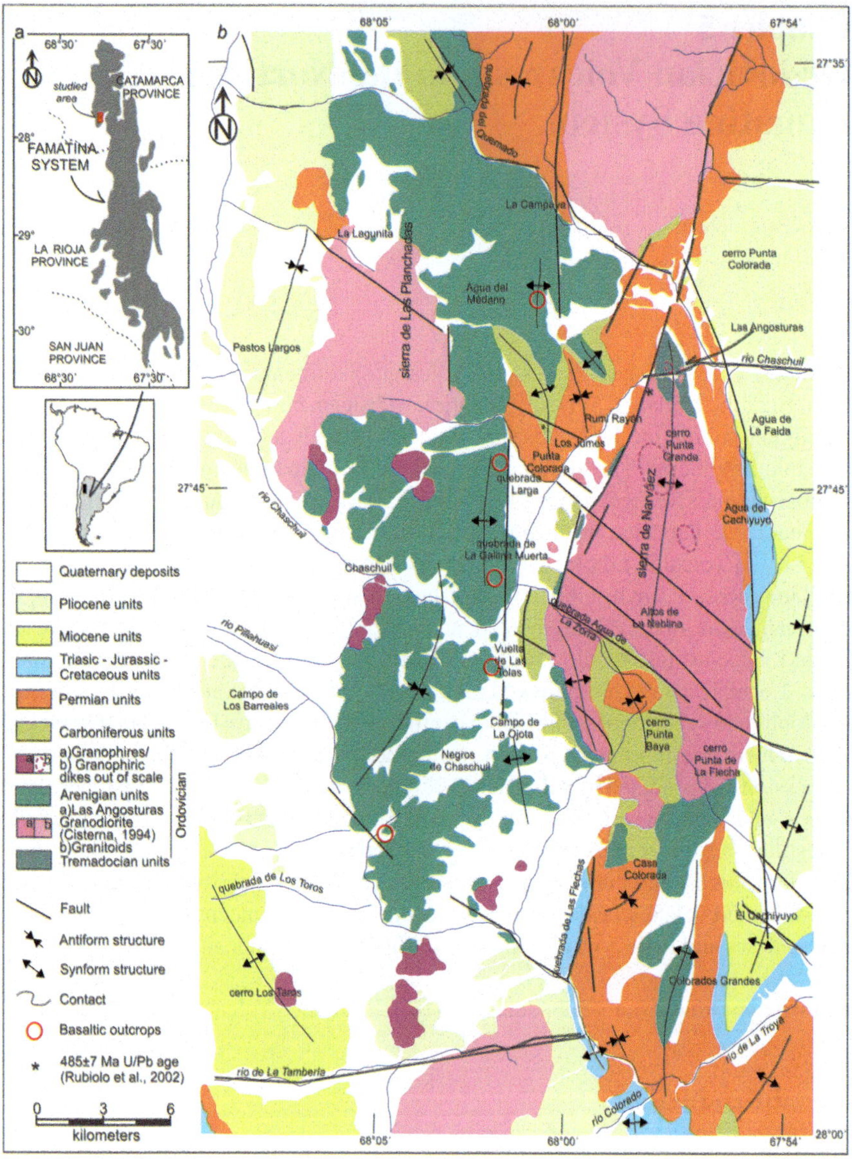

Fig. 5.1 **a** Geological setting of the Famatina System, NW Argentina. **b** Geological map of the north of Famatina System showing the Ordovician volcanic–sedimentary deposits distribution

The Arenigian units (Suri—Las Planchadas Volcanic—Sedimentary Complex, according to Cisterna and Coira 2017), being the most extensive and representative Ordovician deposits in the region, are the focus of analysis at this chapter due they show good examples of submarine and subaerial ancient volcanic successions.

They comprise deposits with a variety of mafic to felsic volcanic lithofacies (Fig. 5.2). The architecture of these units includes mafic to intermediate submarine massive-hyaloclastic flows, felsic lavas/domes, pyroclastic and pyroclastic rich volcaniclastic deposits, intrusions of dykes and rhyolitic cryptodomes. The diversity of volcanic lithofacies recognized in the region mainly reflects variations in processes, environment of emplacement, and eruption styles.

Taking into account the different sections analyzed and other small outcrops along the Planchadas and Narváez hills, the Arenigian stratigraphic succession display approximately 3 km of stratigraphic thickness, showing vertical and lateral variations. The lower portion (1000 m) is dominated by mafic to intermediate volcanic rocks and the upper one is felsic-dominated (2000 m). The first portion of the succession is mainly composed of basaltic to basaltic andesite lavas, hyaloclastites, pillow lavas, pillow breccias and minor dacitic lavas and hyaloclastic equivalents. The upper succession is dominated by volcaniclastic facies, epiclastic and fine-grained volcanogenic sedimentary ones, together with rhyodacitic–rhyolitic and basaltic lavas interlayered among them (Fig. 5.2). The volcaniclastic levels have pyroclastic material components.

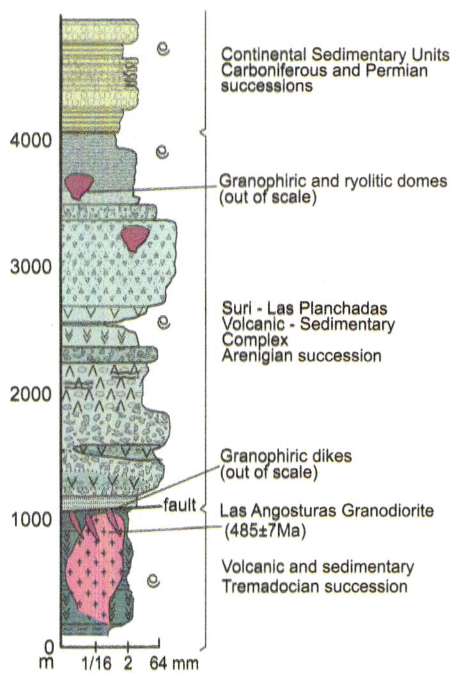

Fig. 5.2 Schematic log showing the internal organization of the Ordovician column

5.2 Basic Lavas and Hyaloclastic Equivalents

The most complete association of basalts and in situ hyaloclastites, grading into breccias of the same composition, that testify the subaqueous emplacement of the lavas, fragmentation processes and reworking of the material developing debris flow deposits, was described along a section of Las Planchadas hill (Figs. 5.1 and 5.3a, b) (Cisterna et al. 2017).

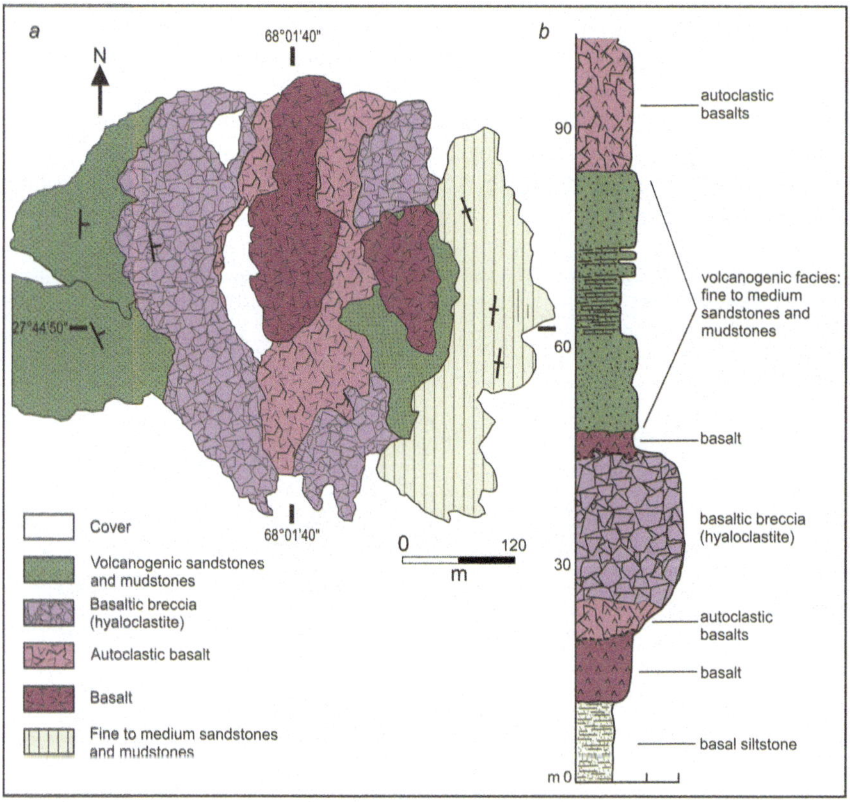

Fig. 5.3 **a** Map of the basaltic volcaniclastic deposits studied. **b** Schematic log showing the internal organization of the analyzed section

The basaltic lavas and minor sills and dykes are mainly present at the lower portions of the complex and the outcrops are discontinuous and extending along a narrow area from north to south, of nearly 25 km long. Their geochemical characteristics, such as LREE enriched REE patterns, enrichment in LILE, in particular K and Pb, and MgO, Fe_2O_3, Ti, Zr, Y, Nb contents indicate a depleted mid-ocean ridge basalt (MORB)-like source, modified by the contribution of subducted components derived from water-rich marine sediments, features compatible with basaltic emplacements in back-arc basins.

The emplacement of basaltic rocks in subaqueous conditions is testified by the association of lava flows, autobreccias, hyaloclastites, and resedimented hyaloclastitic deposits (Cisterna et al. 2017). These lavas are also interbedded with very fine volcanogenic siltstones and fine sandstones developing lava–sediment interaction textures.

Basalts crop out as lenticular or tabular bodies, varying between 2.5 and 10 m thick and usually showing transitional relationships to the autobrecciated facies developing jigsaw-fit textures. The blocky fragments are larger than 3 cm, most of them have curviplanar fracture surfaces and no chilled margins are observed (Fig. 5.4).

Contacts may be sharp or gradational between these deposits and the coherent basalts. The autoclastic basaltic facies, closely related to the massive lavas bodies, are indicative of a nonexplosive fragmentation. They respond to stress in a brittle mode and develop blocks of lava, located at the margins of the lava bodies or dislodged by continuous movement of the flow (cf. McPhie et al. 1993).

Basaltic monomictic breccias facies show clasts shapes that vary between slabby, flow-foliated with jagged ends, ragged or blocky clasts with curviplanar surfaces and display jigsaw structure. Unlike the others facies, clasts develop quenched margins

Fig. 5.4 Autoclastic basalt showing jigsaw texture and a dyke of groundmass rich forming an *in situ* hyaloclastite. Clasts in the hyaloclastite have curviplanar margins and are oxidized. Vesicles are filled by carbonate. Schematic graph of the autoclastic basalt and the hyaloclastite (Hy). (C: clasts, c: crystalloclasts, v: vesicles)

Fig. 5.5 **a** Basaltic monomictic breccia showing slabby clasts. **b** Outcrop of the basaltic monomictic breccias with carbonatic matrix among the clasts and locally, flow-foliation is observed

and the fragment size may vary from less of one mm (millimeter) to 50 cm (centimeters). The clasts show different textures, can be primary flow foliated and have a vitric groundmass or be highly vesicular (Fig. 5.5a, b). These rocks were defined as hyaloclastic basaltic breccias and form deposits up to 60 m thick in transitional or sharp contact related to the autobreccias and massive basalts. Hyaloclastites are composed of clastic aggregates formed by the nonexplosive fracturing and disintegration of quenched lavas (cf. Rittmann 1962). The clasts are formed *in situ* by the intersection of these fractures along the lava flow and by the spalling of quenched glass fragments (cf. McPhie et al. 1993). These clasts may range widely in size from less than one millimeter to tens of centimeters and may be entirely or partly glassy, vesicular or non-vesicular (cf. McPhie et al. 1993) (Fig. 5.6a, b). Autobrecciation and quench fragmentation probably operate together during the disintegration process of the subaqueous lava flow.

Basaltic breccias are composed of clasts showing different structures (Fig. 5.7a, b, e, f), the most frequent have vitric groundmass, sometimes altered to palagonite, with thin plagioclase laths in a pilotaxic groundmass and elongated or subrounded vesicles aligned parallel to the long axis of the clasts (Fig. 5.7a, e), and filled by carbonate or chlorite (Fig. 5.7c). The basaltic clasts may also be rounded and subrounded, with quenched margins and highly vesiculated (Fig. 5.7c). The groundmass is composed

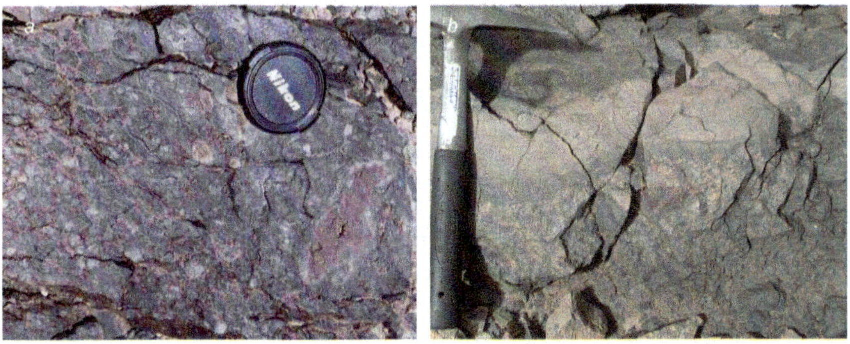

Fig. 5.6 **a** Monomictic basaltic breccia showing clasts with quenched margins and vesiculated. Vesicles are irregular and show flow alignment. The groundmass is composed of glassy particles and phenocryst fragments (plagioclase), carbonate, and palagonite. **b** Contact between the monomictic basaltic breccia (lower sector) and a volcaniclastic breccia with different volcanic clasts components

by glass or oxidized glass, feldspar microlites and vesicles filled by carbonate, chlorite, and quartz; phenocrysts, sometimes fractured, are plagioclase and augite. Plagioclase may be euhedral or display large corroded crystals. They frequently show zonation and abundant glass inclusions. Augite and olivine (as microphenocrysts) are overgrown by amphiboles.

The basaltic breccias described correspond to "*in situ hyaloclastites*" because they display almost all the characteristics indicated for this type of deposits, as clasts with jigsaw texture, pointing that they remained in place while being fragmented (cf. McPhie et al. 1993).

The poorly sorted and quite chaotic nature of the breccia deposits suggests that they are proximal and formed next to its source. Furthermore, the occurrence of angular and fusiform blocks, with their glassy rinds frequently preserved, indicate that breccia deposits were syn-eruptive, preferentially formed in the same space opened up in the lava flow, sites where temporary vents cause the hydromagmatic activity (cf. Carr and Jones 2001).

The mixture of fragments, either massive and angular or highly vesiculated and with glass rinds, may be related to hydromagmatic episodes chiefly due to water–magma interaction, that relieved the pressure on the underlying melt, leading then to the sudden unmixing of the volatile component and to consequent magmatic fragmentation (cf. Dellino and La Volpe 1995).

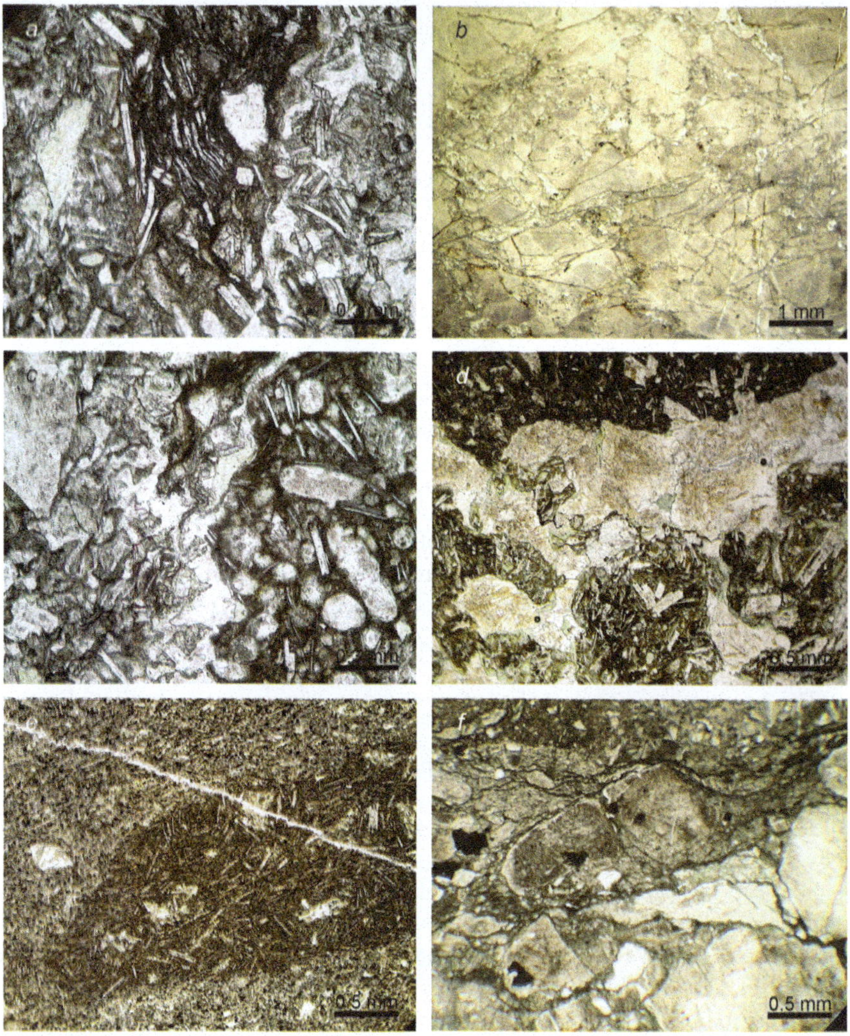

Fig. 5.7 Basaltic hyaloclastic breccias showing **a** An elongated clast with jagged ends in a flow-foliated matrix. **b** Fragments of basalt developing jigsaw structure. **c** Basaltic clast with abundant rounded vesicles filled by carbonate, chlorite, and quartz. **d** Rounded basaltic clast with quenched margins. Calcite is abundant in the groundmass. **e** Fragment of basalt with vitric matrix within a basaltic groundmass. **f** Volcaniclastic basaltic breccia. Plane polarized light in all photomicrographs

5.3 Volcaniclastic Records

The volcanic Arenigian section developed at the northern portion of Famatina System displays a great variety of associated volcaniclastic and volcanogenic sedimentary lithofacies and minor effusive records (Fig. 5.8). The succession is characterized in its basal level by a dacitic lavic member, succeeded by volcaniclastic breccias, sandstones, mudstones, and siltstones bearing abundant lavic and pyroclastic components.

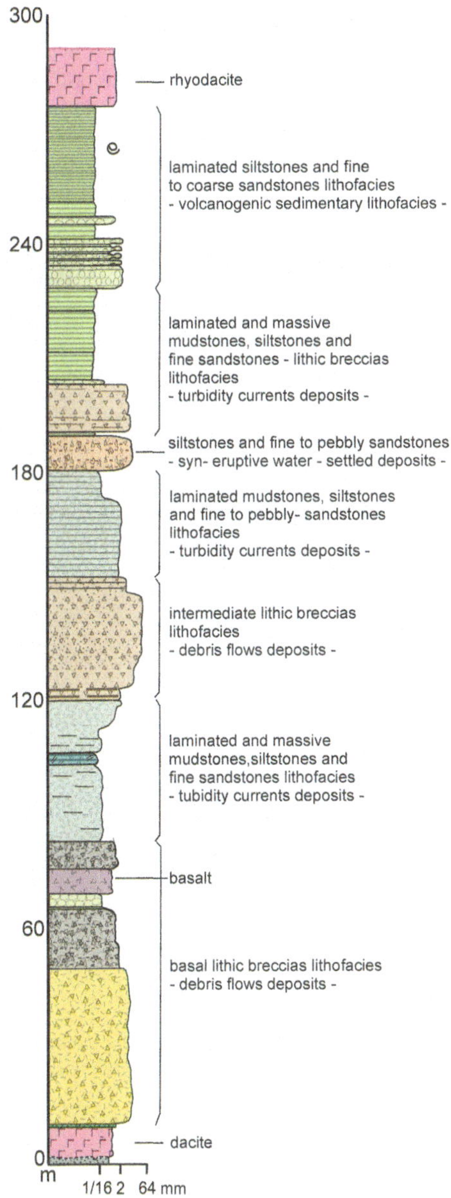

Fig. 5.8 Profile showing the distribution of the volcaniclastic and effusive members along the Arenigian column

These members are immediately followed by volcanogenic sedimentary lithofacies with fossiliferous levels. Among the volcaniclastic lithofacies, there are two main dominant components: volcanic lithic fragments mainly composed of dacite and minor andesite and basalt, together with crystal fragments of quartz and plagioclase and pyroclasts represented by vesicular pumice, glass shard, crystals, and accretionary lapilli. The distinctive characteristics of the different lithofacies recognized give indications of provenance of the components and their origin, types of transport, depositional processes, and particularly environmental conditions of the basin.

The dominant volcaniclastic records are lithic breccias characterized by the volcanic nature of the lithoclast components as well by crystaloclasts. Breccias are poorly sorted and matrix supported, with large lithoclasts up to 30 cm. They can reach 75 m thick and commonly are intercalated with layers of crystal-rich mudstone and sandstone (0.5–2 m thick) (Fig. 5.9a). The lithic components are angular to subangular dacites, andesites, and minor basalts; welded ignimbrites and/or lithic and

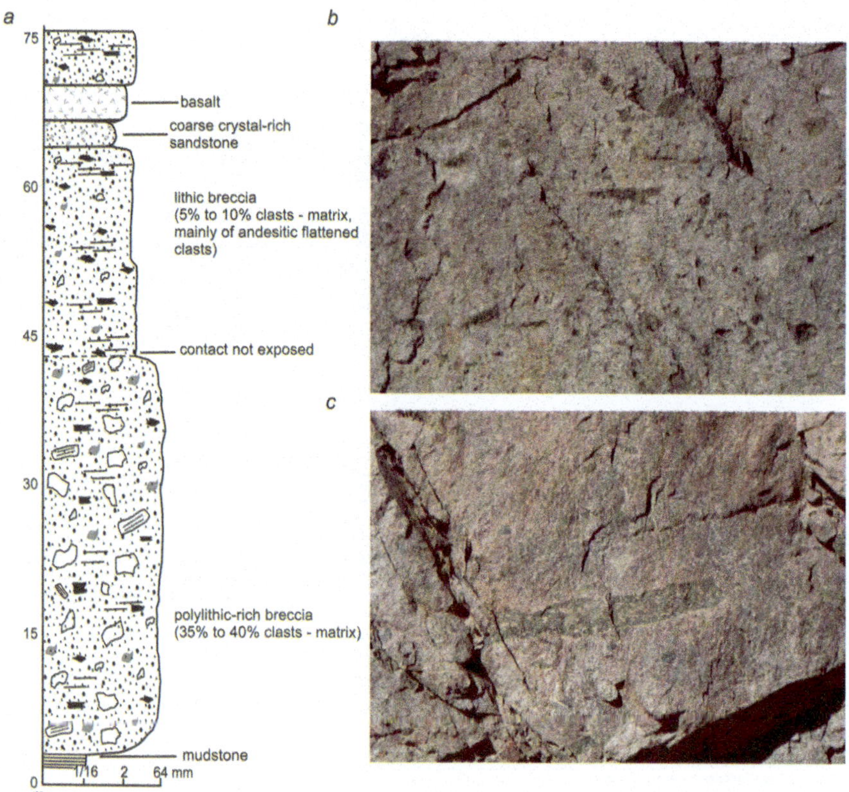

Fig. 5.9 **a** Schematic log showing the internal organization of the basal lithic breccia lithofacies (see Fig. 5.8). **b** Massive lithic breccias with aligned andesitic lithic clasts. **c** Detail showing an angular andesitic clast

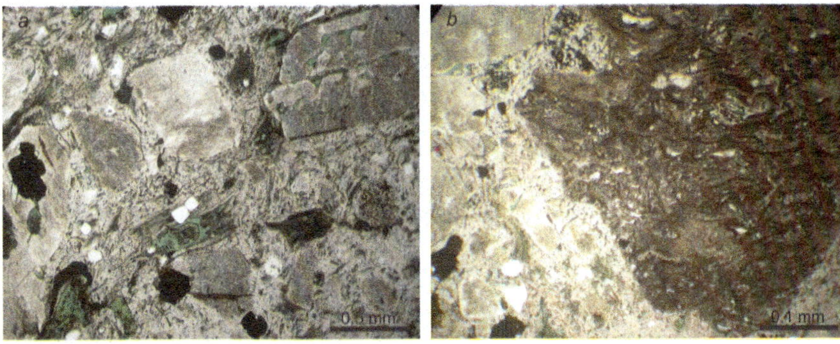

Fig. 5.10 Massive lithic breccias lithofacies showing **a** vitroclastic matrix with platy and cuspate shards contorted and deformed at the margins of dacitic lava lithic clasts and quartz crystal fragments. Plane polarized light. **b** Lithic fragments of ignimbrite are immersed in a vitroclastic rich matrix, where plagioclase and quartz fragmented crystals are also distributed. Plane polarized light

crystallolithic tuffs; and minor laminated mudstone intraclasts (Fig. 5.9b, c). In thin sections can be identified an unsorted matrix, with glass shards and a poor foliation locally developed.

Lithic breccias show changes related to their texture, composition of the lithic fragments (e.g., only of andesitic composition) and the clast-matrix ratio (Fig. 5.10a, b). The presence of both types of components is indicative of the onset of explosive activity at same time of emission of lavas and degradation of flanks of volcanic edifices. The lithic and crystal enrichment related to vitroclastic components, observed overall the dominant lithofacies, point out an effective hydraulic sorting during settling in the water column and during contemporaneous subaqueous resedimentation by mass flow. These are crystal or lithic rich, but show no evidence of having been hot at the time of deposition. So, deposits have not been proven that were formed from gas-supported pyroclastic flows.

The type of transport and depositional processes of these rocks are also reflected through the characteristics of the volcaniclastic lithofacies, most of them are massive and poorly sorted deposits with sharp contacts, showing graded or diffusely bedded structures. Bedforms, tabular and with long laterally development are indicators of rapid (commonly mass flow) deposition. That features characterized water supported mass flows submarine deposits, and according to different authors (Carey and Sigurdsson 1984; McPhie and Allen 1992; Allen et al. 2007) the main types of water supported mass flows in volcanic terranes are turbidity currents, cohesive debris flows, and density modified grain flows. So that, the lithic breccia lithofacies were interpreted as volcaniclastic debris-flow deposits (Cisterna and Coira 2014), indicating high efficiency of the fragmentation processes, strongly controlled in their transport by mass flow processes.

Along this subaqueous succession, there is no evidence of hot gas-supported mode of emplacement. The juvenile nature of the components of breccia deposits and the large thickness of the beds (several meters) are certainly indicators that they are the products of contemporaneous explosive and lavic eruptions, accompanied by sloughing of debris from the flanks of active volcanoes.

Lithofacies composed by volcaniclastic mudstones, siltstones and minor sandstones, massive and laminated, with vitroclastic components, are forming tabular beds that reach up to 45 m thick (Fig. 5.11a). Crystalloclasts are plagioclase fragments and lithoclasts are angular to subangular volcanic clasts (andesites, dacites, and rhyolites), other components are accretionary lapillies and glass shards in the

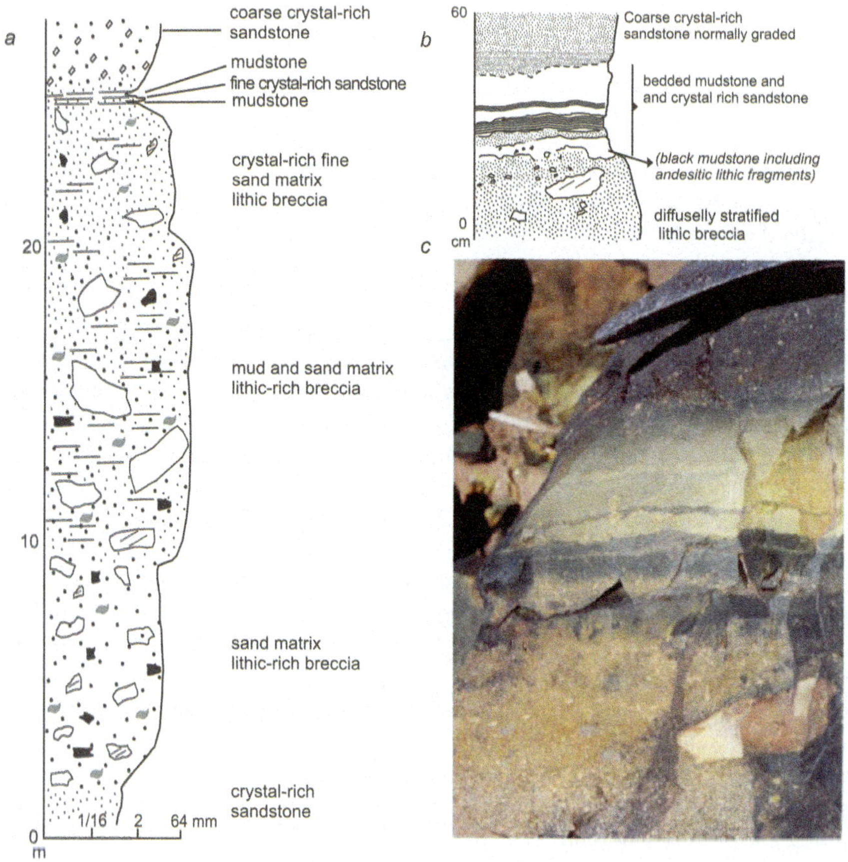

Fig. 5.11 **a** Schematic log of the intermediate lithic breccia lithofacies (see Fig. 5.8) showing the volcaniclastic debris flow internal organization. **b** Detailed log of the upper section. **c** Photograph of a typical subaqueous volcaniclastic debris flow deposit in the same area showing a poorly sorted breccia with angular to subangular lithoclasts (5–15%) of dacites and andesites, 1–10 cm in length, in a massive sandy matrix-support. To the top, are finer beds showing lamination and graded stratification

matrix. They are interstratified with parallel beds of poorly-sorted pebbly sandstones (30 cm thick). Top of the succession predominate siltstones and fine to pebbly sandstones (nearly 10 m thick) where the abundance of pyroclastic materials (crystal fragments, glass shards, and lapillies) and minor lavic fragments is remarkable, in recrystallized vitroclastic matrix (Fig. 5.11b, c). They were defined as *tuff turbidity currents deposits* (Cisterna and Coira 2014).

In volcanic terranes turbidity currents, cohesive debris flows and density modified grain flows are the most important types of mass flow deposits. Particularly common are mass flowage processes which are largely related to gravity instabilities acting on the primary forming facies. By the Arenigian Volcaniclastic facies, the juvenile nature of the components of the breccia records and the large thickness of the beds (several meters) are certainly indicators that they are the products of contemporaneous explosive and lavic eruptions, accompanied by sloughing of debris from the flanks of active volcanoes. Their variations related to the textural and compositional characteristics are the reflex of oscillations in the environmental conditions e.g., enrichment in dacitic and andesitic clasts reflects the proximity of an intra-basin mesosilicic lavic centers; or the abundance of juvenile pyroclastic components related to an explosive eruption from shallow submarine or subaerial vent nearby. On the same basis, facies of laminated and massive mudstones, siltstones, and/or fine to pebbly sandstones (or fine lithic breccias) were recognized as water supported mass flows deposits and compared with turbidity currents (Cisterna and Coira 2014) (Fig. 5.12a, b).

Volcaniclastic facies represented by siltstones and fine to pebbly sandstones was defined as resediment syn-eruptive volcaniclastic deposits (according to the McPhie et al. 1993 criterion) and adjudicated to water settled deposits entirely composed by pyroclastic products (Cisterna and Coira 2014) (Fig. 5.13). These lithofacies are characterized by high content of pyroclasts and the beds have an internally graded and massive structure. The significant volume of shard in these beds could indicate that deposition was relatively rapid, before ocean currents redistributed the material (c.f. Stewart and McPhie 2004).

Synthesizing, the characteristics of the Arenigian volcaniclastic deposits such as massive and poorly sorted texture, sharp contacts, graded or diffusely bedded structures, tabular bedforms and large laterally developed, are indicators of rapid (commonly by mass flow) deposition and testified water supported mass flows submarine deposition. Their composition, characterized by both types of components (volcanic lithic fragments together with crystal fragments and pyroclasts) is indicative of the onset of explosive activity at the same time of emission of lavas and degradation of flanks of volcanic edifices. The lithic and crystal enrichment

Fig. 5.12 **a** Schematic log of the tuff turbiditic deposits showing in **b** mudstones, siltstones and fine sandstones with fluid escape structures. **c** Subaqueous volcaniclastic debris flow deposit showing subangular to angular andesitic and dacitic lithoclasts (1–8 cm in diameter) in an abundant (75–80%) fine sandy matrix

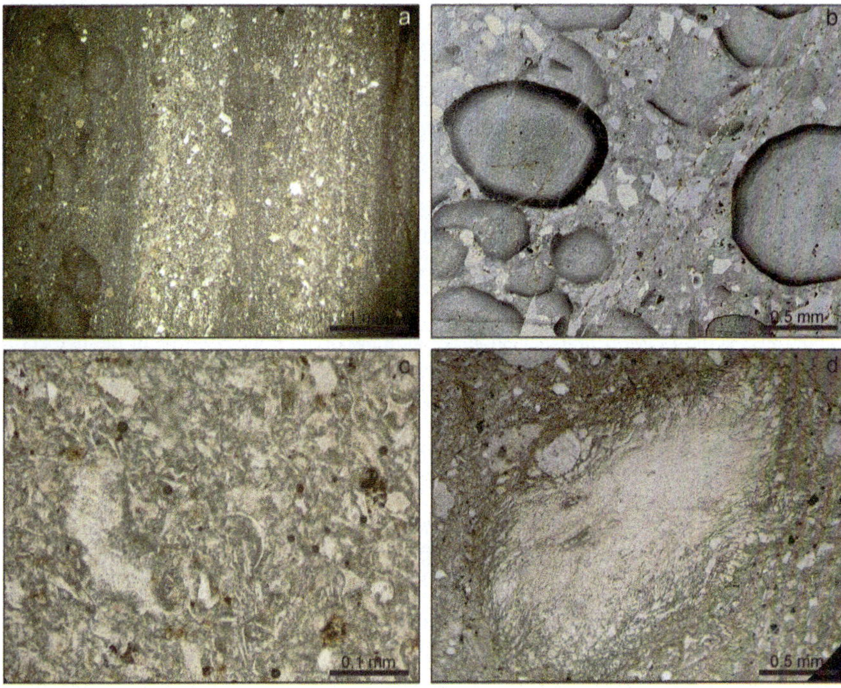

Fig. 5.13 **a** Ash rich turbidite showing fine lamination and zoned accretionary lapillies and **b** detailed view of ash lapillies. Plane polarized light. **c** Vitroclastic matrix with bubble-wall shards showing undeformed cuspate and platy shapes and fine ash filling the interstices. Presence of shards within the matrix of massive siltstones levels. Plane polarized light. **d** Pumice clasts stretched, with internal vesicular microstructures still discernible included in a vitroclastic matrix. Plane polarized light

related to vitroclastic components, observed overall the dominant lithofacies, point out an effective hydraulic sorting during settling in the water column and during contemporaneous subaqueous resedimentation by mass flow (Cisterna and Coira 2014).

References

Allen SR, Hayward BW, Mathews E (2007) A facies model for a submarine volcaniclastic apron: the Miocene Manukau Subgroup, New Zealand. Geol Soc Am Bull 119(5–6):725–742

Carey S, Sigurdsson H (1984) A model of volcanogenic sedimentation in marginal basins. In Kokelaar BP, Howells MF (ed) Marginal basin geology: volcanic and associated sedimentary and tectonic processes in modern and ancient marginal basins. Geological Society [London] Special Publication 16, pp 37–58

Carr PF, Jones BG (2001) The influence of palaeoenvironment and lava flux on the emplacement of submarine, near-shore Late Permian basalt lavas, Sydney Basin (Australia). J Volcanol Geoth Res 112:247–266

Cisterna CE, Coira B (2014) Subaqueous eruption-fed mass-flow deposits: records of the Ordovician arc volcanism in the Northern Famatina Belt; Northwestern Argentina. J S Am Earth Sci 49:73–84

Cisterna CE, Coira B (2017) Registros volcánicos del magmatismo ordovícico en las provincias de Catamarca y La Rioja, noroeste de Argentina. Herramientas para la reconstrucción del arco famatiniano. En Muruaga C, Grosse P (eds) Ciencias de la Tierra y Recursos Naturales del NOA. Relatorio 20 Congreso Geológico Argentino. Tucumán, pp 414–433. ISBN: 978-987-42-6666-8

Cisterna CE, Koukharsky M, Coira B, Günter C, Ulbrich H (2017) Arenigian tholeiitic basalts in the Famatina Ordovician basin, northwestern Argentina: emplacement conditions and their tectonic significance. Andean Geol 44(2):123–146

Dellino PL, Volpe I (1995) Fragmentation versus transportation mechanisms in the pyroclastic sequence of Monte Pilato-Rocche Rosse (Lipari, Italy). J Volcanol Geoth Res 64:211–231

McPhie J, Allen RL (1992) Facies architecture of mineralized submarine volcanic sequences: Cambrian Mount Read Volcanics, western Tasmania. J Econ Geol 87:587–596

McPhie J, Doyle M, Allen R (1993) Volcanic textures: a guide to the interpretation of textures in volcanic rocks. CODES Key Centre, University of Tasmania, Hobart. Tasmania, 198 p

Rittmann A (1962) Volcanoes and their activity. John Wiley and Sons, New York, 305 p

Stewart AL, McPhie J (2004) An Upper Pliocene coarse pumice breccia generated by a shallow submarine explosive eruption Milos, Greece. Bull Volcanol 66:15–28

Chapter 6
Ordovician Volcanism in the Puna

Abstract The orogenic evolution of the Famatinian Cycle in the Puna began during the Upper Cambrian–Lower Tremadocian with records of volcanism conforming two sub-meridian magmatic belts. The Western Magmatic Belt in the Lower Tremadoc is represented by dacitic to rhyolitic lavas, hyaloclastites, tuffs and pyroclastic flows, and basaltic lavas associated with volcaniclastic sandstones, coquinas, and cherts. The Arenigian sequences are characterized by dacitic to rhyolitic pyroclastic flows, lapillites, and tuffs intercalated in volcaniclastic turbiditic sequences. They are generally preceded by basaltic to andesitic massive to pillow lavas and hyaloclastite breccias. The Eastern Magmatic Belt registers begin at Late Cambrian and Upper Tremadoc with dacites and basalts to microgabros syn-sedimentary with the clastic sequences with facies of autobreccias and peperites to the roofs of dacite bodies. During the Tremadoc to Early-Middle Arenig volcanism reaches its climax with siliceous syn-sedimentary lava domes and sills with autoclastic breccias, hyaloclastites and minor spilitized basaltic pillow lavas and lacoliths and sills of microgabros. In both belts geochemical characteristics indicate during Tremadoc early stages of back-arc basins, in Early-Middle Arenigian arc to back-arc on continental crust signature is observed in the Western Magmatic Belt and intraplate signature in mafic rocks and cortical filiations in the siliceous rock of the Eastern Magmatic Belt.

Keywords Ordovician · Volcanism · Puna

6.1 Introduction

The orogenic evolution of the Famatinian Cycle in the Puna began during the Upper Cambrian–Lower Tremadocian with the first records of volcanism conforming two sub-meridian magmatic belts (Western and Eastern Magmatic belts). Such magmatic belts have been defined and interpreted under different approaches, assigning them diverse geotectonic settings, which contemplate in their evolution the collision of allochthonous terranes with the southwest margin of Gondwana or assume parauthochthonous or authochthonous origins (Coira et al. 1982; Allmendinger et al. 1982; Ramos et al. 1984; Dalziel and Forsythe 1985; Ramos 1986; Bahlburg 1990;

B. L. L. Coira and C. E. Cisterna, *Textures, Structures and Processes of Volcanic Successions*, Springer Earth System Sciences,
https://doi.org/10.1007/978-3-030-52010-6_6

Bahlburg and Breitkreuz 1991; Conti et al. 1996). In that way such belts have been interpreted, as associated with two magmatic arcs related to subduction (Dalziel and Forsythe 1985; Ramos 1988; Rapela et al. 1992), genesis proposed in some cases only to the occidental belt (Bahlburg 1990, 1993). In opposition Davidson and Mpodozis (in Coira et al. 1982) and Aceñolaza and Toselli (1984) assign the magmatism of the Eastern Magmatic Belt to an extensional ensialic region developed between the massif of Arequipa to the west and the Brazilian craton to the east and Damm et al. (1990) recognize an intrusive syntectonic arc in the Eastern Magmatic belt, which is associated with the Ocloyic collision. Coira et al. (1999), on the other hand, propose an evolutive model with initial development of an arc, in coincidence with the Western Magmatic Belt (Upper Cambrian–Lower Arenigian), stage that is succeeded by an oblique convergent regime to which a strike-slip system associates in the northeast, and along which the emplacement of mantelic magmas takes place in association with important continental crust fusion processes, whereas in the south an arc magmatism occurs. On this basis and with new geochemical characteristics determined by Coira et al. (2008) that indicate arc and modified MORB like signatures in the case of the volcanism in the Western Magmatic Belt during the Tremadoc, it is proposed for this Belt an early stages of a back-arc basin, where spreading was incipient. Meanwhile the Eastern Magmatic Belt volcanism shows E-MORB characteristics which gives signals of the beginning of an extensional regime, with the development of a back-arc basin with oceanic crust (Coira 2008, 2017; Coira et al. 2009).

During the Early to Middle Arenig, the volcanism in the Western Magmatic Belt shows characteristics of arc to back-arc, developed on a thin continental crust. Meanwhile, in the Eastern Magmatic Belt during the Upper Tremadocian–Early-Middle Arenig the bimodal magmatism (basic rocks with intraplate signature and siliceous rocks with cortical filiation) is indicative of a tectonic environment , where the alkaline magmatism is the result of decompression of the mantle in an extensional environment, while taking place an important cortical fusion. The passage to a compressive regime in the Middle-Upper Ordovician, with which culminates the Ocloyic orogeny, has determined the cessation of volcanism (Coira 2008, 2017).

The Western Magmatic belt extends in Puna toward the west in Chile (Fig. 6.1), it includes volcano-sedimentary sequences with trilobite faunas of Lower Tremadoc age (Moya et al. 1993) as well as others with graptofaunas Arenig s.l. to Middle-Late Arenig age and with less precision assignable to Early Llanvirn (Monteros et al. 1996). In the Lower Tremadoc sequences, the volcanism is represented at the west of Salar del Rincón (Vega de Lari and Pinato) and S Cerro Rincón by dacitic to rhyolitic lavas, hyaloclastitas, tuffs and pyroclastic flows, and basaltic lavas (Koukharsky et al. 1996) associated with sandstones volcaniclastic, siltstones, coquinas, and cherts. The Arenigian–Early Llanvirn sequences are characterized by dacitic to rhyolitic pyroclastic flows, lapillites, and tuffs intercalated in volcaniclastic turbiditic sequences as is recorded in the Sierra de Guayaos (Koukharsky et al. 1988), Vega Lari (Koukharsky et al. 1996), E and SE of the Salar de Pocitos (Blasco and Zappettini 1996), Salina de Jama (Coira and Nullo 1989), and Agua de la Perdiz (Breitkeuz 1986) in Chile. They are generally preceded by basaltic to andesitic massive lavas, pillow lavas and

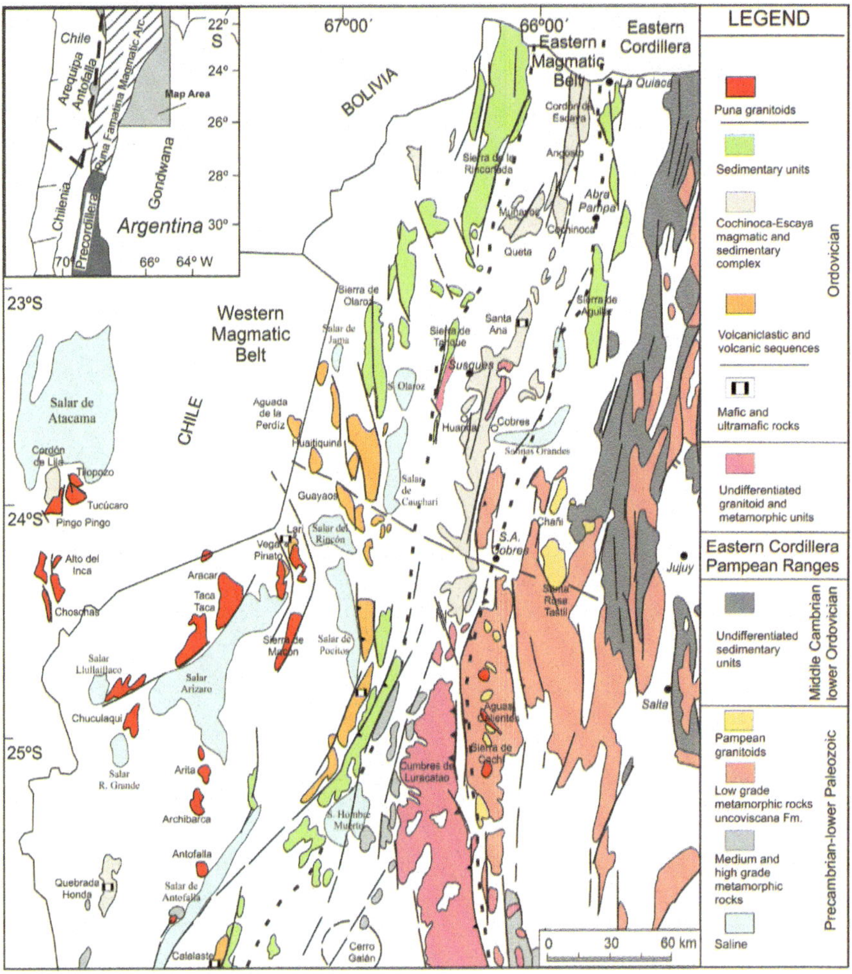

Fig. 6.1 Map of part of northern Argentina and Chile showing the distribution of Ordovician and temporally related magmatic and sedimentary units. Inset shows the Puna–Famatina Magmatic Arc and Map area (modified from Coira et al. 1999)

pillow fragmented breccias and hyaloclastite breccias (see Huaitiquina type section, Coira and Barber 1989, Fig. 6.2).

The Eastern Magmatic Belt registers, in the northern Puna (22°–24°30′S) begins in at the end of the Cambrian and beginning of the Tremadoc with bimodal character, represented by dacites and basalts to microgabros syn-sedimentary with the clastic host sequences.

Sills of dacitic composition outcropping in the Cerro Niño Muerto (SW Salinas Grandes), dated in 495 ± 4 Ma (Hauser et al. 2008), present to their roofs levels of peperites in the dacite/metasediments interface (Coira et al. 2008). Toward the

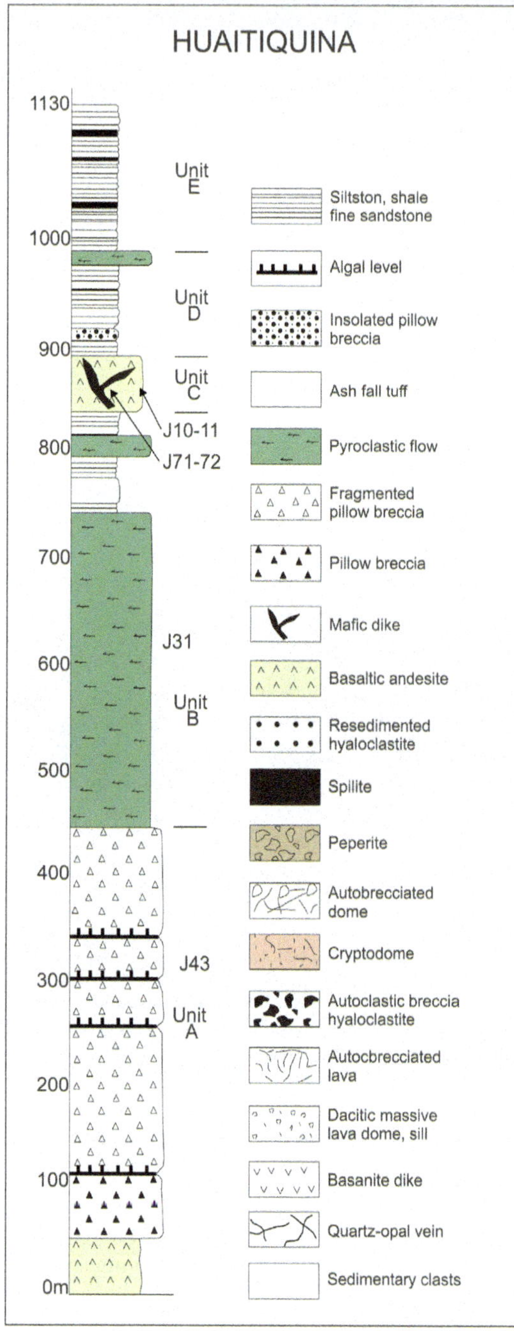

Fig. 6.2 Type section in the Huaitiquina river (from Coira and Barber 1989)

southern sections, at NE of Salar Centenario, outcrop metasediments associated with deformed meta-rhyolites and dacites dated in 485 ± 5 Ma and metabasalts with a mineral association indicative of greenschist facies metamorphic conditions (Viramonte et al. 2007). Other records of this stage of volcanism in the Sierra de Cobre and Sierra de Tanque are the basalts and microgabros that integrate the sedimentary-magmatic sequence that hosted the granite plutons of Cobres and Tanque (475–480 Ma) (Lork and Bahlburg 1993; Perez and Coira 1998; Haschke et al. 2005; Coira 2008). Those basalts and microgabros conform bodies concordant with the hosting sediments, which are deformed, frequently presenting lenticular and rosary structures.

During the Tremadoc to Early-Middle Arenig volcanism in the the Eastern Magmatic Belt reached its highest representation with siliceous syn-sedimentary lava domes and sills with autoclastic breccias, hyaloclastites frequently observed to their margins and minor spilitized basaltic pillow lavas as well as microgabros forming lacoliths and sills (see Cochinoca-Escaya Magmatic-Sedimentary Complex, example Fig. 6.3). (Coira and Koukharsky 1991; Coira et al.1999). Granitoids with U/Pb 476 Ma-444 Ma ages (Lork et al. 1991; Lork and Bahlburg, 1993; Bahlburg et al. 2016) are also part of the Eastern Magmatic Belt.

The deformation and low-grade metamorphism that affect the magmatic-sedimentary sequences of the Ordovician in the Puna along both belts make it difficult to identify the original textures of the magmatic rocks, as of their primary relationships with the associated sedimentary facies, particularly taking into account the contrasting rheologies among them.

In the following chapter, it considers the analysis of those characteristic facies of the Ordovician volcanism of the Puna, along Western and Eastern Magmatic belts which well represented and exposed show distinctive features, whose observation and detail analysis provides tools to be applied to reconstruct eruptive and depositional processes in ancient volcanic successions.

6.2 Textural-Structural Characteristics and Emplacement Conditions

The deformation and low-grade metamorphism that affect the magmatic-sedimentary sequences of the Lower Paleozoic in the region make it difficult to identify the original textures of the magmatic rocks, as of their primary relationships with the associated sedimentary facies, particularly taking into account the contrasting rheologies among them.

In the following presentation, it considers the analysis of those characteristic facies of the Lower Paleozoic volcanism of the Puna, which well represented and exposed show distinctive features, whose observation and detail analysis provides tools to be applied to reconstruct eruptive and depositional processes in ancient volcanic successions.

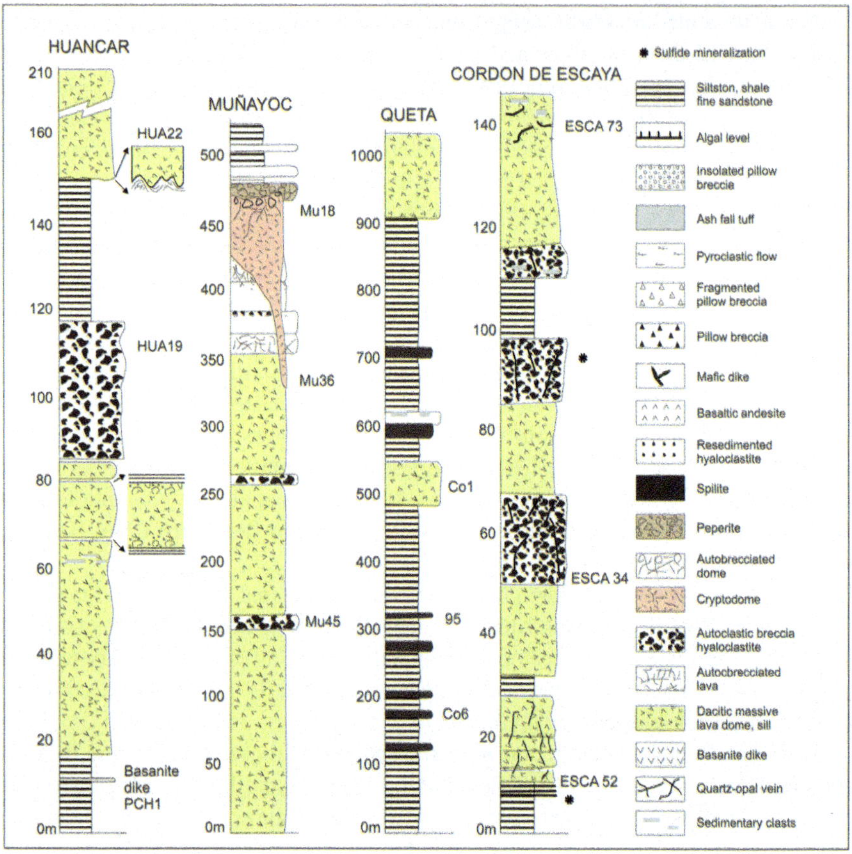

Fig. 6.3 Stratigraphic sections showing facies of representative sections of the Cochinoca–Escaya Magmatic–Sedimentary Complex (Coira et al. 1999, 2004) of the Eastern Magmatic Belt

6.2.1 Eastern Magmatic Belt Facies

Detailed studies in sectors of lower deformation of the Eastern Magmatic Belt have allowed to recognize, within the dominant siliceous facies, textures, and structures that testify submarine to subaqueous conditions during the evolution of the volcanism in that Belt that are important for description and interpretation of that processes. Dacitic lavas, domes, and sills together with hyaloclastic and peperitic levels (Coira and Koukharsky 1994; Coira and Pérez 2002), as well as spilitic lavas with pillow structure interspersed in the turbiditic sequences characterize that Belt (Coira and Koukharsky 1991), represented by the Cochinoca-Escaya Magmatic and Sedimentary Complex (Coira et al. 2004).

The lavas, domes, and sills are composed of porphyritic dacites with 20–35% phenocrysts of oligoclase quartz, biotite, and megacrysts of alkali feldspar up 5 cm across in a devitrified microgranular groundmass that contains aggregates of quartz, alkali feldspar, chlorite, and sericite (Fig. 6.4a). These rocks use to show deformation (Fig. 6.4b). Perlitic textures use to be recognized in these rocks reflecting hydration and cooling processes (Fig. 6.5a–d).

Further fragmentation by cooling of the perlitic dacites (Fig. 6.6a) determines false glass shards (Fig. 6.6b) that resemble those resulting from explosive episodes of

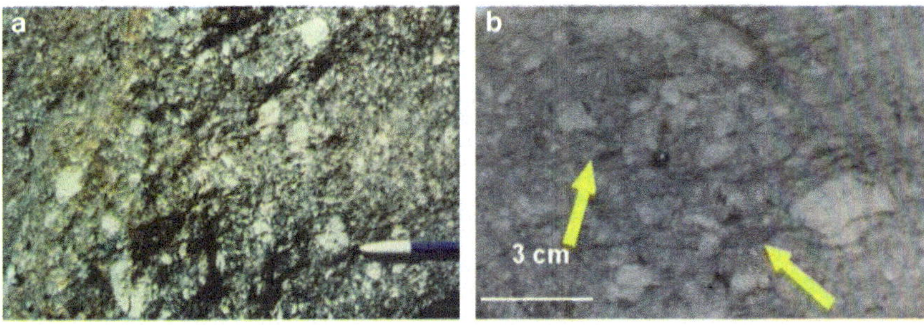

Fig. 6.4 **a** Dacite rich in crystals with alkali feldspar megacrystals (3–5 cm) and phenocrysts of oligoclase, quartz, biotite, **b** Porphyritic dacite with evidence of deformation. Feldspar, quartz, and biotite phenoclasts showing stretching

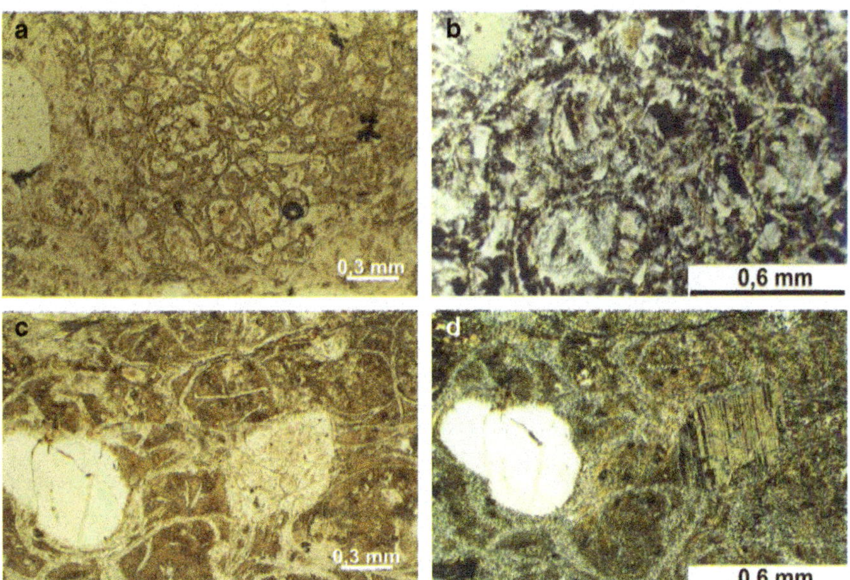

Fig. 6.5 **a** Dacitic lava with perlitic groundmass. Parallel polarized light. **b** Perlitic groundmass devitrified to quartz and feldspar intergrowths. Plane polarized light **c** Perlitic texture of dacite lava. Plane polarized light. **d** Show cracks highlighted by microaggregates of quartz and feldspar. Crossed polarized light

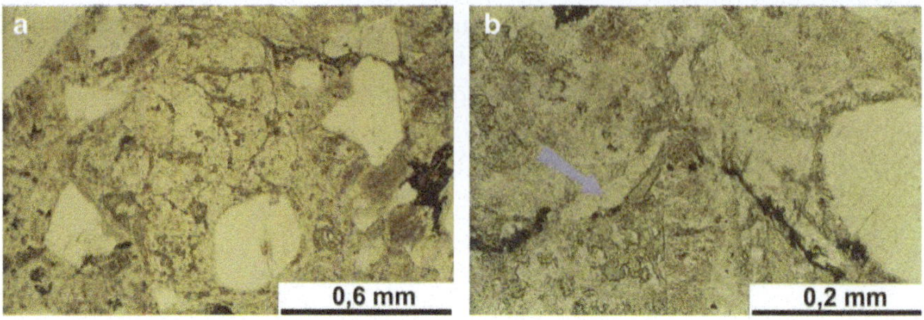

Fig. 6.6 **a** Fragmentation of perlitic lavas by cooling. Plane polarized light. **b** False glass shards result of hyaloclastitization of perlitic glasses pointed with an arrow. Plane polarized light

vesiculated magmas. The dome bodies (40–100 m thick) usually present, on the tops and flanks, autobreccias, hyaloclastitic, and peperitic facies, those that are succeeded by volcaniclastic levels corresponding to resedimented hyaloclastites. The dacitic sills use to develop at their contacts with the sedimentary host rocks, hyaloclastites, and/or peperites.

The autobreccias (tens of meters thick) are formed by irregular dacitic monomictic blocks developing jigsaw-fit textures most of them showing curviplanar fracture surfaces and no chilled margins. They use to be in gradational contact with hyaloclastic facies.

The dacitic hyaloclastic facies developed on tops of domes, with thicknesses of a few meters up to 30 m, are constituted by fragments of the dacite 0.3–1 m

Fig. 6.7 **a** Dacitic hyaloclastite on top of dome with fragments of angular to subangular dacite without evidence of great rotation in moderately rich matrix of the dame composition. **b** Detail of hyaloclastite with dacitic clasts in abundant hyaloclastite matrix

Fig. 6.8 **a** Hyaloclastite with curviplanar to angular phyric dacite clasts in moderately matrix rich finely comminuted of the same composition. **b** Detail of dacitic clasts in a matrix including dacite splinters up to 5 mm in diameter and crystal fragments. **c** Dacite lava margin showing passage to hyaloclastite. Plane polarized light. **d** Hyaloclastic margin of dacitic blocks of laccolith megabreccia. Blocky angular to subangular dacite fragments display closely packed jigsaw-fit texture. In detail, the boundaries of larger clasts are fragmented into microgranules and crystals (in situ microfragmentation. Plane polarized light. **e** Dacitic hyaloclastite with fragmentary crystals of quartz, feldspar and biotite and curviplane devitrified vitroclastos. Crossed polarized light. **f** Fragment of lava with passage to hyaloclastite matrix. Plane polarized light

across, angular to subangular, partly curviplanar, that are arranged with a texture of type "jigsaw," without evidence of greater rotation, immersed in a scarce to moderately rich fine matrix, composed of microfragments, frequent splinters of similar composition (Figs. 2.7a, b and 2.8a–d).

These textures indicate nonexplosive fragmentation processes, product of thermal stress, and as the result of sudden cooling of magma in contact with water or intruded in wet and poorly consolidated sediments. The passage of these facies to the massive

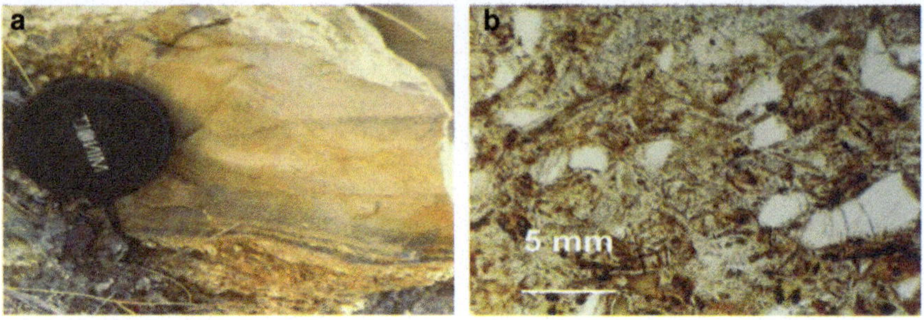

Fig. 6.9 **a** Resedimented hyaloclastite normal graded. **b** Finely granulated resedimented hyaloclastite with equant vitreous fragments and crystals. Plane polarized light

dacites or to the autobreccias is usually gradational. Meanwhile on the flanks of the domes, there are accumulated resedimented hyaloclastites finely granulated with equant vitreous fragments and crystals frequently normal graded (Fig. 6.9a, b).

Diverse types of peperites, from blocky megapeperites to microglobular peperites, are well exposed in several sections of Sierras de Quichagua and Queta (Coira and Pérez 2002) particularly along the margins of dacitic domes (30–100 m thick) or sills (5–8 m across) and lesser extent related to spilitic lavas. Contacts of these rocks related with the sedimentary host rocks are commonly characterized by the development of mixed rocks (igneous and sedimentary) with diverse clastic textures, from blocky shaped igneous clasts which preserve in places a jigsaw-fit arrangement, to lobate or globular magma bodies, both dispersed in a massive and usually indurated sedimentary rock. The mixed rocks grade into the unmixed coherent magmatic bodies and also into the sedimentary host rock, which shows its typical and distinctive laminated structure in contrast with the structureless mudstone matrix in the mixed rock, characteristics that fit the definition of peperite (White et al. 2000). Examples of the blocky peperite facies (see sketch of blocky megapeperite above jigsaw dacite breccia on top of a dome, Fig. 6.10b) are represented by levels of large dacitic clasts, tens of meters thick, set in a sedimentary massive matrix, at the top of a dacitic dome which grade from the dome by an increase in clast separation and sediment content. The coherent dacitic dome (up to 80 m thick) gradually changes toward the roof into a matrix-poor in situ hyaloclastite (8–10 m thick) with jigsaw-fit to slightly rotated subangular partly curviplanar fragments 0.5–1 m across. Along the side of those hyaloclastic dome facies, angular to subangular dacitic fragments, 0.5–3 m across form a close-packed breccia level, 4 m thick (Fig. 6.10a-a). Upwards, closer to the sedimentary host rock they are slightly displaced with respect to one another and separated by minor silicified silty host rock. The close-packed breccia is gradually succeeded by a more loosely packed breccia 6 m thick (Fig. 6.10a-b) characterized by subangular dacitic blocks, 0.2–1.2 m across, some with partly rounded and spalled margins enclosed in a structureless indurated silicified siltstone matrix (Fig. 6.10c).

Fig. 6.10 Megapeperites with blocky textures on top of a dome: **a** closely packed; **b** unpacked. **b** Sketch of blocky megapeperite above jigsaw dacite breccia on top of a dome. **c** Close-up showing dispersion of subangular dacitic fragments into the sediments, adjacent to large dacite blocks

Details on the contact magma-sediment enable to identify angular microfragments and crystal chips dispersed in a slightly modified sedimentary matrix near the contact (Fig. 6.11a, b).

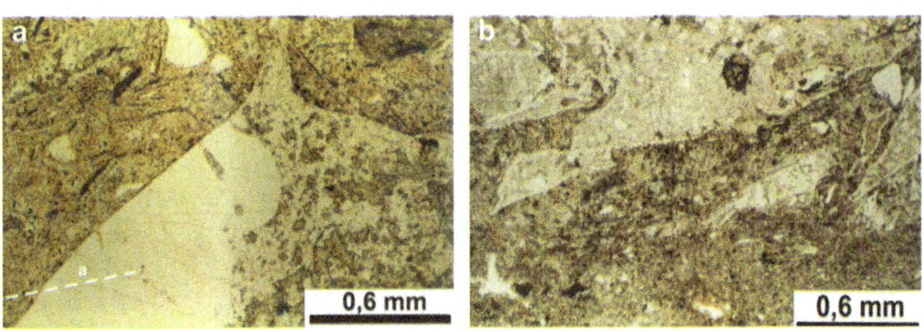

Fig. 6.11 **a** Detail of peperite showing a dacitic clast, fragments of crystals, and minor chips of the same composition in a siltstone matrix. Plane polarized light **b** Ragged dacite clast margin of dacite megabreccia. Angular dacitic microfragments and crystal chips are dispersed in a slightly silicified sedimentary matrix near the dacite contact. Plane polarized light

The blocks are distributed in a way that suggests detachment of dacitic fragments from the dome and dispersion into the sedimentary host rock, which is mainly structureless siltstone that shows lamination away from the contact. The main processes inferred for these deposits, on the basis of their textures and facies arrange, are quench fragmentation of magma on contact with wet sediments, and dynamic stressing of earlier chilled body margins by continued injection of magmas, acting together in many cases with fluidization of the wet sediments controlled by pressure reduction during opening of cooling cracks in the body (Coira and Pérez 2002). It is possible to infer in addition to the previous processes, minor explosive expansion of water incorporated into the system (bulk interaction steam explosivity, Kokelaar 1986).

The globular peperites (according to the Busby-Spera and White 1987 criterion) also identified in the margins of some of the dacitic laccoliths, domes, or sills are characterized by matrix-supported globular clasts. Examples at the top and base of the sills (15–20 m thick) show layers of 0.03–0.10 m thick of matrix-supported (50–60% in volume) deposits with globular clasts. (see sketch on Fig. 6.12a). Dacitic globules, comprising drop-like or bulbous fragments with irregular outlines and varying in size from 2 to 15 mm in diameter, are isolated within the indurated siltstone host rock

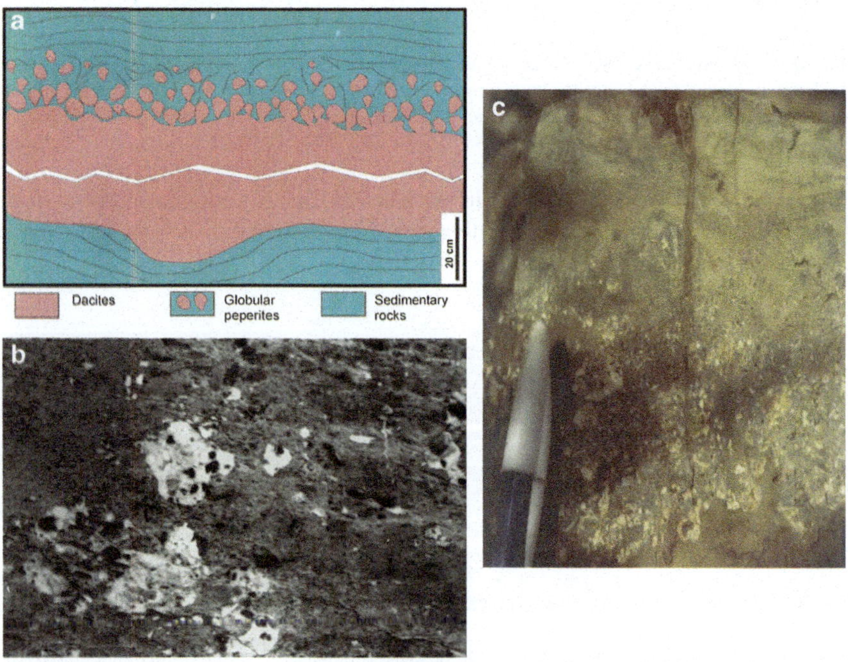

Fig. 6.12 **a** Sketch of globular peperites on the roof of a dacitic sill. **b** Dacite globules and equant fragments of variable size detached from the dacite sill and dispersed in the mudstone host rock. **c** Microglobular dacitic fragments and crystals are dispersed within the texturally homogeneous mudstone. Note the distinct loss of lamination of the sedimentary host rock near the small projection of the dacitic intrusion

(Fig. 6.12b). Near the dacitic sill contact, the sedimentary matrix is homogeneous and structureless. Vesicles, vuggy cavities 150–400 μm and irregular channels are identified in the matrix at the contact with the dacite (Coira and Pérez 2002). These globular peperites show textural signs of fluidization processes (Fig. 6.12c), with evidence of passive emplacement of magma without hydroclastic fragmentation.

The magma and host showed fluidal behavior, with detachment of globules of magma as a result of oscillation of vapor films along the magma/wet sediment interfaces and accompanied by bulk sediment transport away from the contacts. This texture of peperite suggests, in contraposition to blocky peperites described, a deeper emplacement beneath sea level where confining pressure was enough to suppress steam explosions, but not so high as to prevent the achievement of a certain minimum vapor flow velocity required for fluidization (Kokelaar 1982). The sketch (Fig. 6.13) indicated the control of confining pressure on the development of blocky versus globular peperites.

Sills of dacitic composition, up to 50 m thick, outcropping in the Cerro Niño Muerto (SW Salinas Grandes, see Fig. 6.1) also present to their roofs levels of peperites in the dacite/metasediments interface. In this case, fragments of angular to

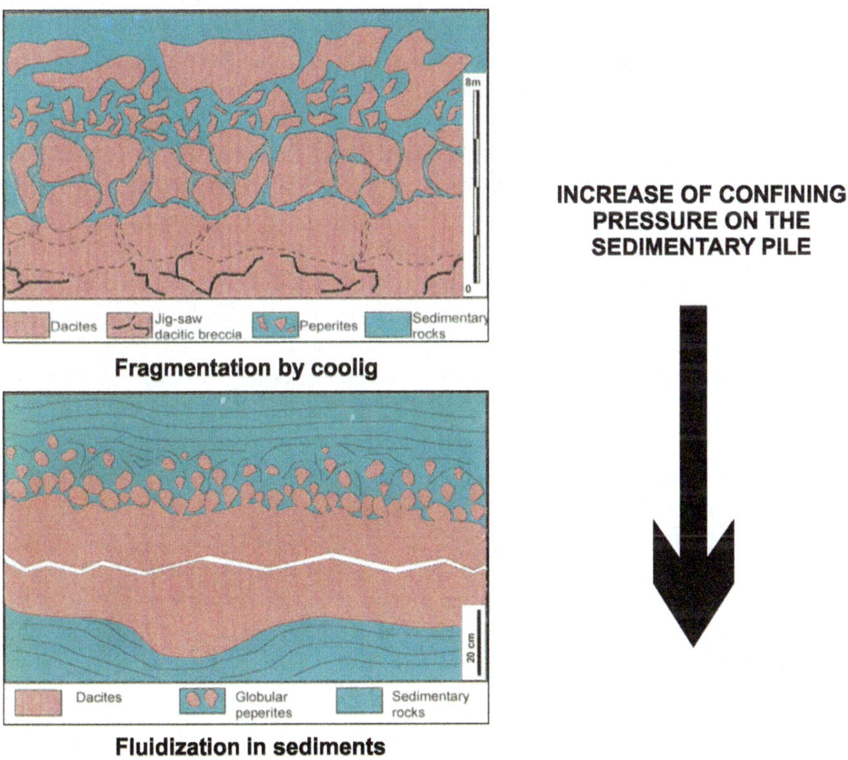

Fig. 6.13 Sketch showing the control of confining pressure on the development of blocky versus globular peperites

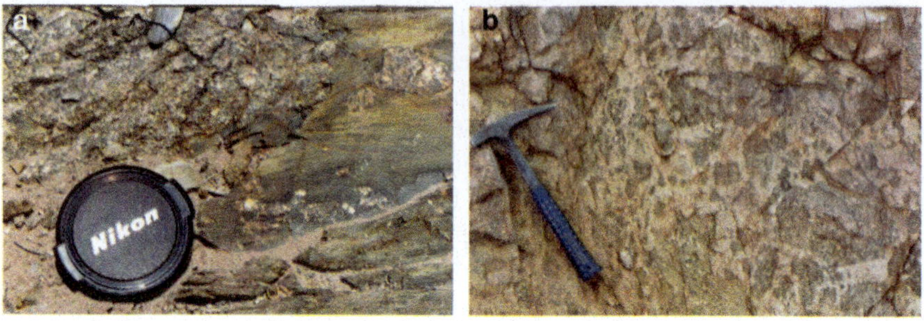

Fig. 6.14 **a** Blocky peperite. Fragments of angular to subangular dacites, 1 cm up to 20 cm in diameter, are included in a structureless indurated psamytic-pelitic matrix. They show a certain degree of elongation as a result of superimposition of low-grade dynamothermic metamorphism. **b** Loosely packed dacite breccia at the interface dacite/peperite

subangular dacites, (blocky shaped), 1–20 cm in diameter, are included in a structureless indurated silicified psamytic-pelitic matrix, with evidence of fluidization and loss of its original texture (Fig. 6.14a) indicating quench fragmentation of magma on contact with wet sediments, and dynamic stressing of chilled body margins by continued injection of magmas testifying the syn-sedimentary nature of these dacites. The passage of the dacite to the peperite is gradual through facies of close-packed autobreccias gradually succeeded by a more loosely packed breccia (Fig. 6.14b) (Coira et al. 2008).

The pillowed spilitic lavas are characterized by their lobed structure, with evidence of elongation, resulting in superimposed deformation and development of zonal arrangements (Coira and Koukharsky 1991). Well expose section outcrops at Cordón de Escaya where they conform layers, twenty to twenty-five meters thick of interlaced pillows with ellipsoidal contours and diameters of 1 × 0.4 m–3 × 1.5 m, closely packed (Fig. 6.15a). The pillows show a cooled margin poorly porphyritic (1–2 cm thick) with abundant vesicles (15–20% in volume) filled with quartz and chlorite. The intermediate zone (15–40 cm thick), more porphyritic, is characterized by the development of concentric zones with contrasting vesiculation (3–15% in volume) replaced by chlorite, carbonate, and quartz. The core does not present zonation and is characterized by elongated vesicles, irregular, and with strong development of carbonates, chlorite, and quartz.

The spilite sills also show toward the roof and at the contact with the sedimentary host rock the development of blocky peperites where spilite clasts with irregular margins and slight displacement from one another are immersed in a indurated mudstone (Fig. 6.15b). Those sills also show, toward the roof, fractures filled with siliceous mudstone host rock denoting the interaction magma wet sediments (Fig. 6.15c).

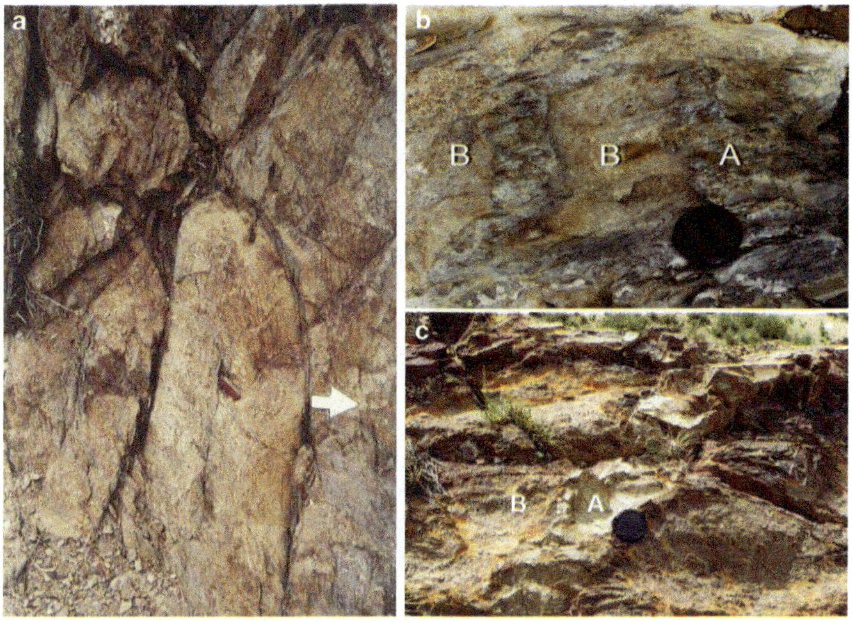

Fig. 6.15 **a** Pillowed spilitic lavas with lobed structure, closely packed, with evidence of superimposed deformation. Oblique section to the direction of flow. The arrow indicates the roof of the section. **b** Blocky breccia on top of spilite sill. Mudstone (**A**) enclosing spilite clasts (**B**) which show irregular margins and slight displacement from one another. **c** Roof of spilite (**B**) sill with a siliceous mudstone filled fracture. Note the partly irregular form of the fracture and homogeneous nature of the sediment enclosed in it (**A**)

6.2.2 Western Magmatic Belt Facies

The presentation of the textural and structural characteristics of Western Magmatic Belt facies is not comprehensive, rather it emphasizes on the analysis of those facies that show the more distinctive textures and structures. This is the case of well-exposed basaltic to andesitic successions of the Arenigian volcano-sedimentary sequences that outcrop in the Huaitiquina River (Coira and Barber 1989). They generally precede the explosive dacitic to rhyolitic stage with minor lavas and are normally arranged from base to top as follows: massive lavas, pillow lavas, pillow breccias, and hyaloclastite breccias.

The basaltic–andesitic pillow lavas are roughly elliptical bodies, with pillow-like shapes characteristic of interconnected tubes and lobes of lava (McPhie et al. 1993) elongated reflecting some degree of superimposed deformation. They are poorly (Fig. 6.16a) to moderately porphyritic, constituted by plagioclase tablets without preferential orientation to which crystals of clinopyroxene are associated, both distributed in a mesostasis rich in chlorite and opaque minerals, showing in some cases albitization of plagioclase (Fig. 6.16b). Their margins are cooled with

Fig. 6.16 **a** Basaltic–andesite poorly porphyritic with a vesiculate groundmass rich in chlorite and opaque minerals. Plane polarized light. **b** Basaltic–andesitic lavas-albitized plagioclase, augite in a mesostasis of peninite-clinochlore, titanite, and opaque. Crossed polarized light. **c** Basaltic–andesite pillow breccia with broken pillow lobes spalled from pillow lavas. **d** Detail of basaltic–andesite clasts with cooled margins in the pillow breccia. **e** Hyaloclastite with jigsaw fit to slightly rotated dacite blocks in a moderately matrix rich. **f** Siliceous ashfall characterized by vitroclastic texture with bubble walls and tubular glass shards. Plane polarized light

scarce phenocrysts and abundant vesicles. The abundant vesicles allow to infer an emplacement in a submarine environment at depth < 500 m (Moore and Schilling 1973).

Pillow breccias (Carlisle 1963; Studigel and Schmincke 1984) comprise broken pillow lobes, joint bounded blocks, and granules spalled from pillow rinds. They are monomitic and include a mixture of clast derived from different parts of pillow lobes (Fig. 6.16c, d). The hyaloclastic breccias consist of whole pillows, fragments of pillows, or parts of thin lava flows, within a tuff matrix of devitrified and metasomatized microglobular particles and shard-like fragments. The components of the matrix were not derived from pillow or flow margins but formed by micro-globulation and shattering.

Among the siliceous members of the Western Magmatic Belt facies usually are recognized dacitic to rhyolitic sills (20–30 m thick) being conformable bodies with respect to the host sedimentary deposits. It observed a gradual transition from the massive dacitic or rhyolitic bodies, internally coherent, not vitroclastic, and uniformly porphyritic into hyaloclastites layers along its roof (2–5 m thick) and with minor development at their base.

These hyaloclastites use to show blocky monomictic clasts (Fig. 6.16e) immersed in a microgranular matrix composed of microfragments frequent splinters of similar composition and sporadically distributed crystal fragments. These characteristics indicate nonexplosive fragmentation processes and thermal stress associated with cooling of the magma intruded in wet sediments.

The siliceous ashfall deposits dacitic to rhyolitic, may be homogenous and lack internal structure or have a normal graded stratification and in some cases reverse stratification in beds of 2 up to 10 m thick. The tephras are characterized by vitroclastic textures and in the best-preserved sections, is possible to recognize shards 0.10–0.40 mm in size with bubble walls and ovoid or tubular and filamentous vesicles (Fig. 6.16f) in some cases showing welding between them. Usually, they are replaced by quartz–feldspar microaggregates as the fine mesostasis. In some cases, a few pumice fragments (0.7–1 mm in diameter) are present. The crystals, scarce (2–5%) to abundant (20–30%) are generally angular and composed of andesine, oligo-andesine, quartz, and biotite. These deposits use to be associated with volcaniclastic levels, with evidence of syn-eruptive resedimentation (according to McPhie et al. 1993).

References

Aceñolaza F, Toselli A (1984) Lower Ordovician volcanism in northwest Argentina. In: Brunton DL (ed) Aspects in the Ordovician system: Paleontological Contributions of the University of Oslo, vol 295, pp 203–209

Allmendinger R, Jordan T, Palma M, Ramos VA (1982) Perfil estructural en la Puna catamarqueña (25–27°S), Argentina. Actas del 5° Congreso Latinoamericano de Geología 1:499–518. Buenos Aires

Breitkreuz C (1986) Das Paläozoikum in dem Kordillere Nordcchiles (21°-25°S). Geotekton Forsch 70:1–88

Bahlburg H (1990) The Ordovician Puna basin in the Puna in the NW Argentina and N Chile: geodynamic evolution from back-arc to foreland basin. Geotekton Forsch 75:1–107

Balhburg H (1993) Hypothetical southeast Pacific continent revisited: new evidence from middle Paleozoic basins of northern Chile. Geology 21:909–912

Bahlburg H, Breitkreuz C (1991) The evolution of marginal basins in the southern central Andes of Argentina and Chile during the Paleozoic. J S Am Earth Sci 4:171–188

Bahlburg H, Berndt J, Gerdes A (2016) The ages and tectonic setting of the Faja Eruptiva de la Puna Oriental, Ordovician, NW Argentina. Lithos 256–257:41–54

Blasco G, Zappettini E (1996) Boletín 217. Programa Nacional de Cartas Geologicas. Servicio Geologico Minero Argentino. Hoja geológica San Antonio de los Cobres 2566-1, escala 1:250.000
Busby-Spera CJ, White JDL (1987) Variation in peperite textures associated with differing hostsediment properties. Bull Volcanol 49:765–775
Carlisle D (1963) Pillow breccias and their aquagene tuffs: Quadra Island, British Columbia. Am J Sci 71:48–71
Coira B (2008) Volcanismo del Paleozoico inferior en la Puna Jujeña. In: Coira B, Zappettini EO (eds) Geología y Recursos Naturales de la Provincia de Jujuy, Relatorio del XVII Congreso Geológico Argentino. Asociación Geológica Argentina. Buenos Aires, pp. 140–154
Coira B (2017) Volcanismo Paleozoico de Salta y Jujuy. In: Muruaga C, Grosse P (eds) Ciencias de la Tierra y Recursos Naturales del NOA. Relatorio del XX Congreso Geológico Argentino. Asociación Geológica Argentina. San Miguel de Tucumán, pp 410–423, ISBN 978-1-873671-00-9
Coira B, Barber E (1989) Vulcanismo submarino ordovícico (Arenigiano-Llanvirniano) del Río Huaitiquina, Provincia de Salta, Argentina. Rev de la Asociación Geológica Argentina 44:68–77
Coira B, Nullo F (1989) Facies piroclásticas del volcanismo ordovícico (Arenigiano-Llanvirniano) Salina de Jama, Jujuy. Rev Asoc Geol Argentina XLIV(1–4):89–95
Coira B, Koukharsky M (1991) Lavas en almohadilla ordovícicas en el Cordón de Escaya, Puna septentrional, Argentina. Actas 6° Congreso Geológico Chileno 1:674–678
Coira B, Koukharsky M (1994) Complejos Submarinos Dómico-Lávicos Silíceos de Edad Ordovícica en el Sector Oriental de Puna Jujeña (22–23°45'S). Sus Implicancias. Actas 7° Congreso Geológico Chileno 2:1000–1004
Coira B, Pérez B (2002) Peperitic textures of Ordovician dacitic synsedimentary intrusions in Argentina's Puna Highland: clues to emplacement conditions. J Volcanol Geotherm Res 14:165–180
Coira B, Davidson J, Mpodozis C, Ramos V (1982) Tectonic and magmatic evolution of the Andes Northern Argentina and Chile. Earth Sci Rev 18:303–332
Coira BL, Mahlburg Kay S, Pérez B, Woll B, Hanning M, Flores P (1999) Magmatic sources and tectonic setting of Gondwana margin Ordovician magmas, northern Puna of Argentina and Chile. In: Ramos VA, Keppie JD (eds) Laurentia-Gondwana Connections before Pangea. Geological Society of America Special Paper 336. Boulder, Colorado, pp 145–170
Coira B, Caffe P, Ramírez A, Chayle W, Díaz A, Rosas S, Pérez A, Pérez B, Orozco O, Martínez M (2004) Descripción de la Hoja Geológica 2366–1 "Mina Pirquitas." Boletín del Servicio Geológico Minero Argentino N° 269:1–123
Coira B, Díaz AM, Cisterna C (2008) Magmatismo sin-sedimentario Cámbrico tardío/Tremadociano en el Cerro Niño Muerto y su asignación a la faja magmática oriental de la Puna. Actas del XVII Congreso Geológico Argentino I:195–196
Coira B, Koukharsky M, Ribeiro Guevara S, Cisterna CE (2009) Puna (Argentina) and Northern Chile Ordovician basic magmatism: A contribution to the tectonic setting. J S Am Earth Sci 27(1):24–35
Conti CM, Rapalini AE, Coira B, Koukharsky M (1996) Paleomagnetic evidence of an early Paleozoic rotated terrane in northwest Argentina. A clue for Gondwana-Laurentia interaction? Geology 24(10):953–956. Boulder, Co, USA
Dalziel IWD, Forsythe RD (1985) Andean evolution and the terrane concept. In: Howell DG (ed) Tectono-stratigraphic terranes of the CircumPacific Region: Circum-Pacific-Council for Energy and Mineral Resources Earth Science Series, vol 1, pp 565–581
Damm KW, Pichowiak S, Harmon RS, Todt W, Kelley S, Omarini R, Niemeyer H (1990) Pre-Mesozoic evolution of the central Andes; The basement revisited. In: Kay SM, Rapela CW (eds) Plutonism from Antarctica to Alaska. Geological Society of America Special Paper 241. Boulder, Colorado, pp 101–126
Haschke M, Deeken A, Insel N, Grove M, Smichtt A (2005) Growth pattern of the Andean Puna plateau constrained by apatite (U/Th)/He, K-feldspar 40Ar/39Ar, and zircon U-Pb geochronology.

In: 6th international symposium on andean geodynamics-ISAG 2005. Extended Abstracts. Barcelona, pp 360–363

Hauser N, Matteini M, Pimentel MM, Omarini RH (2008) Petrology and La–ICPMS U–Pb Geochronology of volcanic rocks of the Lower paleozoic rocks units of the Central Andes, NW Argentina: implications for the evolution of western Gondwana. In: VI Symposio Sudamerican de Geología Isotópica, San Carlos de Bariloche, Argentina, Proceedings CD-rom

Kokelaar BP (1982) Fluidization of wet sediments during emplacement and cooling of various igneous rocks. J Geol Soc Lond 139:21–33

Kokelaar BP (1986) Magma-water interactions in subaqueous and emergent basaltic volcanism. Bull Volcanol 48:275–289

Koukharsky M, Coira B, Morello O (1988) Vulcanismo ordovícico de la Sierra de Guayaos, Puna salteña. Características petrológicas e implicancias tectónicas. Rev Asoc Geol Argentina Tomo XLIII(3):19

Koukharsky M, Torres Claro R, Etcheverría M, Vaccari NE, Waisfeld B (1996) Episodios volcánicos del Tremadociano y del Arenigiano en Vega Pinato, Puna salteña, Argentina. Actas 13° Congreso Geológico Argentino 5:535–542

Lork A, Bahlburg H (1993) Precise U/Pb ages of monazites from the Faja Eruptiva de la Puna oriental and the Cordillera oriental, NW, Argentina. XII Congreso Geológico Argentino, Actas 4:1–6

Lork A, Granet B, Kramm U, Miller H (1991) U-Pb investigations of monazite and polyphase zircon: implications for age and petrogenesis of trondjemita of the Southern Cordillera Oriental NW Argentina. 6° Congreso Geológico Chileno. Acta I:398–402

McPhie J, Doyle M, Allen R (1993) Volcanic textures. A guide to the interpretation of textures in volcanic rocks. Centre for Ore Deposit and Exploration Studies. PhD thesis, University of Tasmania, Tasmania, 198p

Monteros JA, Moya MC, Monaldi CL (1996) Graptofaunas arenigianas en el borde occidental de la Puna Argentina: Implicancias paleogeográficas. Actas 12° Congreso Geológico de Bolivia 2:733–746

Moore JG, Schilling JG (1973) Vesicles, water and sulphur in Reykjanes Ridge basalts. Contrib Mineral Petrol 41:105–118

Moya MC, Malanca S, Hongn F, Bahlburg H (1993) El Tremadoc temprano en la Puna occidental argentina. 12° ° Congreso Geológico Argentino y 2° Congreso de Explotación de Hidrocarburos 2:20–30. Mendoza

Pérez B, Coira B (1998) El magmatismo ordovícico de la Sierra de Tanque, Puna septentrional, Argentina. Memorias XIII Congreso Geológico de Bolivia I:229–235. Potosí, Bolivia

Ramos VA (1986) El diastrofismo oclóyico: un ejemplo de tectónica de colisión durante el Eopaleozoico en el Noroeste Argentino. Rev del Instituto de Geología y Minería. UNJu 6:13–28. Jujuy

Ramos VA (1988) Late Proterozoic-Early Paleozoic of South America: a collisional history. Episodes 11(3):168–175

Ramos VA, Jordan T, Allmendinger R, Kay S, Cortés J, Palma M (1984) Chilenia: un terreno alóctono en la evolución paleozoica de los Andes centrales. 9° Congreso Geológico Argentino. Actas 2:84–106. Bariloche

Rapela CW, Coira B, Toselli A, Saavedra J (1992) The lower Paleozoic magmatism of southwestern Gondwana and the evolution of the Famatinian orogen. Int Geol Rev 34:1081–1142

Studigel H, Schmincke HU (1984) The Pliocene seamount series of La Palma, canary Islands. J. Geophys Res 89:11195–11215

Viramonte JM, Becchio R, Viramonte JG, Pimentel MM, Martino RD (2007) Ordovician igneous and metamorphic units in southeastern Puna: new U-Pb and Sm–Nd data and implications for the evolution of northwestern Argentina. J S Am Earth Sci 24:167–183

White JLD, McPhie J, Skilling I (2000) Peperite: a useful genetic term. Bull Volcanol 62:65–66

Chapter 7
Carboniferous Extensional Volcanism

Abstract Carboniferous volcanic records outcropping along the northwestern of Famatina Belt-southeastern Puna, NW Argentina, display continuous and well-exposed deposits of volcanic rocks, constituted by rhyolites and lesser volume of basalts and trachytes, related with volcaniclastic deposits. These rocks, assigned to the Lower Pennsylvanian (320 Ma U–Pb age), represent a bimodal volcanic succession whose geochemical and isotopic characteristics point to an extensional source regimen. Petrographic, structural, and textural characteristics of the sequence give indications of the components nature, emplacement, and environmental conditions. The association of porphyritic and vitrophyric lavic rhyolites with equivalent breccias is consistent with the eruption mechanism of central lava domes. The dominance of rhyolitic lithic clasts in the breccias suggests explosive eruptions. Basalts, as concordant lava flows, display autoclastic facies at their margins indicating locally high strain rates in an already more brittle lava body during continuous effusion. The association of rhyolites with their volcanoclastic breccias followed by minor basaltic and intermediate lavas suggests episodic eruption pulses triggering instabilities in the volcanic edifices and degradation of their flanks, with emission of lava and breccia flows and concomitant collapse of the central lava domes. The volcanic succession was locally affected by a later shear deformation episode.

Keywords Carboniferous volcanism · Bimodal · Extensional

7.1 Introduction

The northwestern of Famatina System and south of Puna region present volcanic-volcaniclastic records whose ages allow to extend the Early Carboniferous volcanism to the Lower Pennsylvanian in the northwestern border of the Famatinian Belt (Cisterna et al. 2013; Zappettini et al. 2015; Coira et al. 2016).

Late Paleozoic records overlie unconformably, deformed Precambrian to Early Paleozoic metamorphic basement units, as well as Early Ordovician granitoids and volcanic deposits. They are dominantly composed of extensive siliciclastic beds

B. L. L. Coira and C. E. Cisterna, *Textures, Structures and Processes of Volcanic Successions*, Springer Earth System Sciences,
https://doi.org/10.1007/978-3-030-52010-6_7

assigned to the Carboniferous and Permian (Turner 1960; Cuerda 1965). Nevertheless, a Carboniferous volcanic succession was being identified in the last years to the west of the Chaschuil valley (Fig. 7.1), where U–Pb SHRIMP ages of 348 ± 3–342 ± 1 Ma (Martina et al. 2011) and 320 ± 2 Ma U–Pb SHRIMP were reported (Cisterna et al. 2013; Zappettini et al. 2015), the last one obtained from samples of the Las Lozas section (Fig. 7.1). Carboniferous volcanic records represent a bimodal volcanism characterized by the presence of subalkaline/tholeiitic rhyolitic members, together with alkaline basalts and trachytes. The basalts display features comparable to transitional MORB and within-plate tholeiites, with contributions from a mantle source affected by crustal contamination (Coira et al. 2016). The acid members have geochemical affinities to within-plate magmas and compositions suggesting derivation from continental crustal material with mantle source interaction or juvenile essentially mantle-derived crust (Coira et al. 2016).

The geochemical data are consistent with the Hf isotopic geochemistry that indicates a juvenile source of Cryogenian age without evidence of older reworked crust input (Zappettini et al. 2015). The Chaschuil Valley at the northwestern of Famatina System-Southern Puna (Fig. 7.1) display a continuous and well-exposed Carboniferous succession where petrographic, structural, and textural characteristics point out the volcanic processes related to their source and emplacement, taking into account that they are one of the few records of the Late Paleozoic volcanism in northwestern Argentina, in the context of the Proto-Andean margin of Gondwana.

7.2 Textural and Structural Characteristics and Emplacement Conditions

Late Paleozoic units were gently folded and faulted during the Cenozoic (Andean) deformation, nevertheless the volcanic members preserve primary depositional structures (e.g., spherulites, flow bands, banding) and textures, still retain near-glassy (cryptocrystalline) domains and mafic and felsic minerals, including phenocrysts that sometimes appear variably altered to chlorite and sericite. Locally, they were affected by an important ductile shear deformation (Cisterna et al. 2006) (e.g., along the Pastos Largos Valley), developing structural characteristics useful to define this process. The Carboniferous records are characterized by abundance of acid lavas and their fragmented equivalents, together with minor intercalated occurrences of basic to intermediate ones.

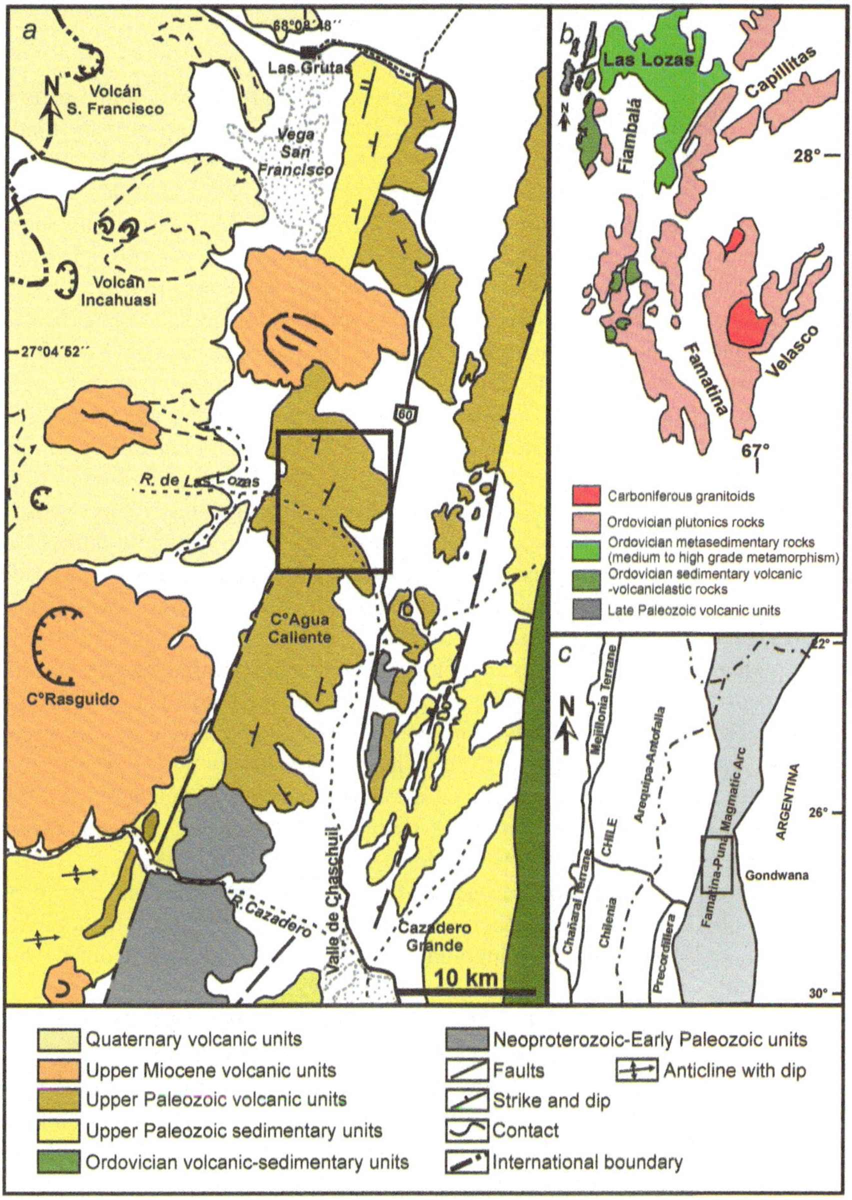

Fig. 7.1 **a** Geological map of Las Lozas Valley showing the location of the studied area; **b** Regional geological sketch comprising Las Lozas region; **c** Geological setting of Puna-Famatina- System, NW Argentina

7.2.1 Effusive Lithofacies

They are mainly composed of rhyolites, represented along the entire column as flows reaching a thickness of about 25 m (Fig. 7.2). Rhyolites develop primary fluidal structures (Fig. 7.3a, b) and are bordered by brecciated monomytic equivalents, both at the top and bottom of the lava flows. They commonly show well development and preserved spherulitic texture (Fig. 7.3c). Rhyolites are porphyritic with plagioclase, K-feldspar, and quartz euhedral megacrysts set in a glassy, partially spherulitic microcrystalline groundmass showing well-developed flow banding (Fig. 7.4a, b). Mafic components are mostly oxidized biotite, small and scarce (<1%), and sometimes fine prismatic disoriented grains of arfvedsonite (Fig. 7.4c, d). Groundmass showing primary banding with devitrification textures, spherulites with microcrystalline fibers of quartz and K-feldspar crystals and micrographic intergrowths, is common (Figs. 7.4e, f and 7.5a) and indicates relatively rapid crystallization of a glassy precursor at high temperatures (cf. Lofgren 1971; Murphy and March 1993; Allen et al. 2008). The formation of spherulitic to micrographic intergrowths is consistent with a process of slower cooling (Lofgren 1971). Rhyolitic vitrophyres exhibit cryptocrystalline or glassy groundmass with flow banding (Fig. 7.5b) occasionally showing vitrophyric layers alternating with others showing devitrification features (fine granophyric intergrowth together with spherulites) (Fig. 7.5c, d).

The coherent rhyolitic lavas transitionally change into fractured autoclastic beds, always poorly sorted, with a thickness up to 15 m, and predominance of angular to subangular rhyolitic clasts immersed in a finer-grained groundmass of acid fragments (Fig. 7.6). The blocky clasts, about 3 cm or more in size, locally show jigsaw texture. This texture is difficult to be recognized where the matrix is abundant (c.f. McPhie et al. 1993). Clasts have curviplanar fractures and non-quenched margins, feature consistent with brecciated margins of a dome (e.g., Allen et al. 2008).

Groundmass of the breccias, display flow-banded structure, with fragments showing spherulitic and granophyric intergrowths but also glassy and perlitic textures (Fig. 7.7a), suggesting a high-temperature environment, active during their development.

Autoclastic rhyolites change into monomictic rhyolitic breccias developing tabular bodies with sharp boundaries, up to 50 m in thickness. They are poorly sorted, clasts are angular to subangular, massive and locally display foliated structure with alignment of the clasts. Sizes of them vary among 1–5 cm, rarely reaching up to 60 cm: clast proportions may vary between 10 and 45% of the total. Clasts are rhyolitic lava fragments, such as porphyritic ones, fragments composed of a fine microcrystalline quartz–feldspar mosaic (Fig. 7.7b); clasts with granophyric or spherulitic textures; sometimes, clasts with flow orientation. The matrix is composed of a siliceous microcrystalline to very fine aggregate, locally showing flow structure.

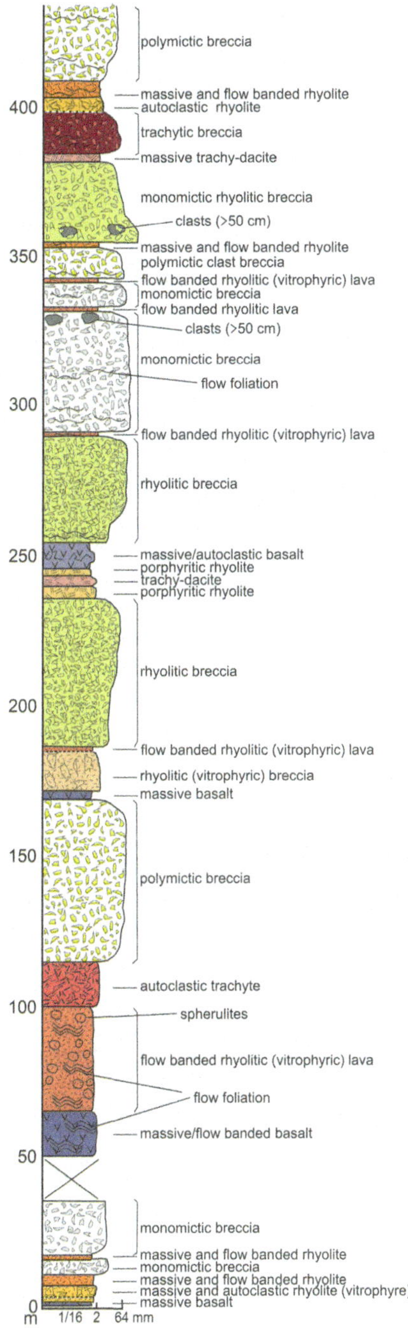

Fig. 7.2 Schematic log showing the internal organization of the Las Lozas section and location of studied samples

Fig. 7.3 **a** Parallel flow-foliated grey rhyolitic lava. **b** Locally, primary fluidal structure shows foliation rotations, indicating plastically deforming parts of the rhyolitic lava flow. **c** Rhyolite showing isolated and coalesced spherulites, spherical and ellipsoidal, up to 3 cm

Trachytic and trachydacitic members of the Carboniferous sequence were developed at the upper levels of the stratigraphic column (e.g., Las Lozas section, Fig. 7.2). They show dark-colors and form beds up to 15 m thick, represented by lava flow mainly autobrecciated with angular clasts ranging from 3 to 10 cm in size. Their texture is porphyritic, with predominance of K-feldspar phenocrysts (10–15%) up to 3 mm in size that show chessboard texture. Mafic minerals are subordinate (subhedral augite and hornblende). The groundmass is greenish-brown, with trachytic, partly glassy, texture with weakly to moderately oriented feldspar laths. Vesicles, filled with iron oxides and quartz, may be aligned parallel to the flow direction. The preexisting glass changes into quartz–feldspar aggregates and/or altered into chlorite. Inclusions of subrounded to irregular lithic fragments of volcanic rocks with a very fine-grained groundmass, partly flow-banded, are also observed under the microscope.

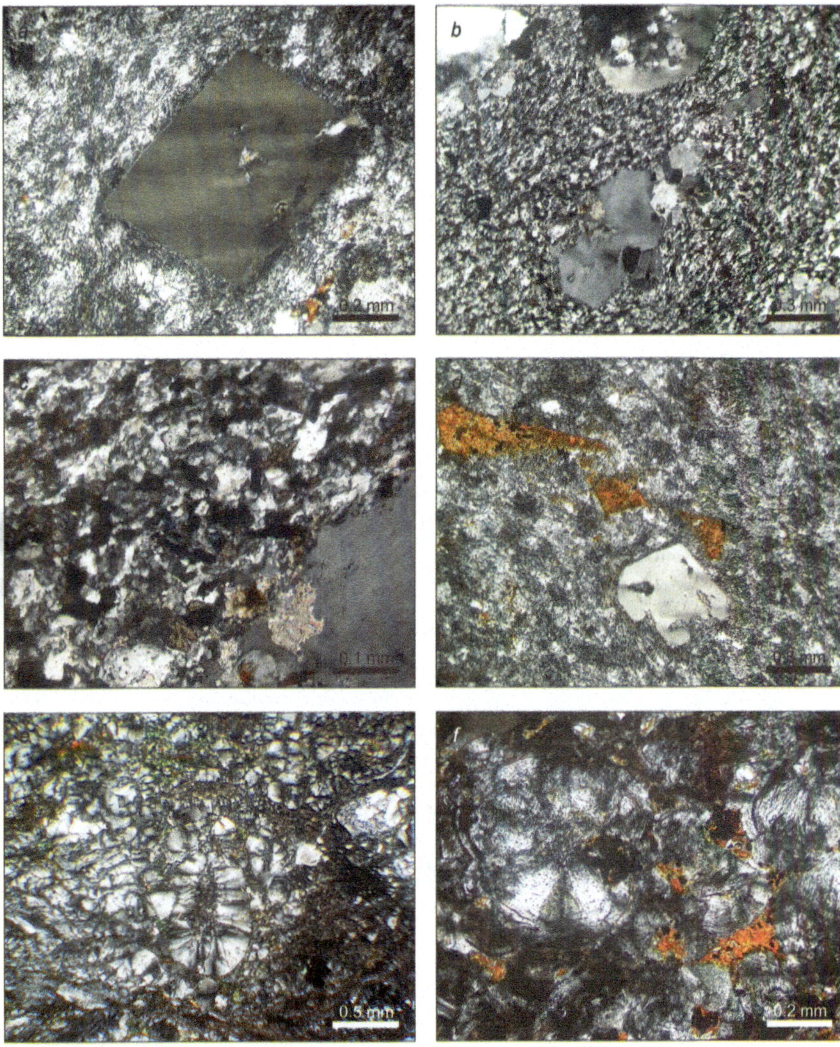

Fig. 7.4 Photomicrographs of rhyolites showing **a** euhedral quartz phenocryst within a cryptocrystalline to glassy groundmass with flow banding. **b** Flow foliated groundmass rich in arfvedsonite, where partially resorbed quartz phenocrysts are oriented according to the main structure. **c** Detailed view of the arfvedsonite groundmass. **d** Quartz phenocryst showing embayments and glass inclusions. Oxidized biotite phenocrysts. **e** Spherulite in a groundmass showing partial recrystallization with quartz and feldspar intergrowths. **f** Spherical isolated spherulite defined by the radiating disposition of quartz and feldspar crystal fibers. They display radial extinction. (Crossed polarized light)

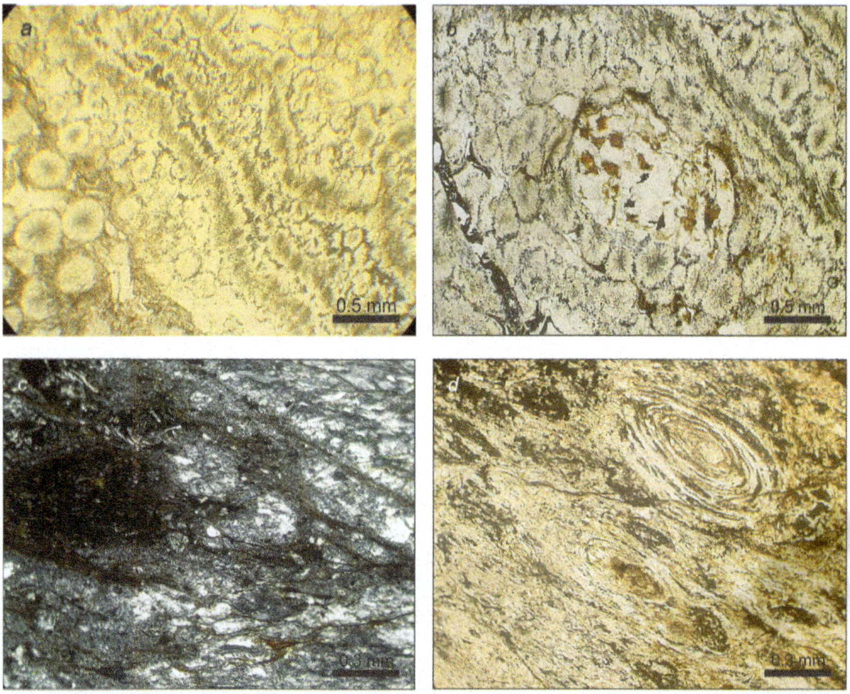

Fig. 7.5 Photomicrographs showing: **a** and **b** rhyolite showing flow-banded structure defined by the spherulitic arrangement in trains. In the center of the picture, quartz phenocryst is surrounded by spherulites. (Parallel polarized light). **c** Vitrophyric groundmass of a finely microcrystalline quartz–feldspar mosaic developing layer structure. (Crossed polarized light). **d** Perlitic cracks defining concentrically arrangement around subspherical cores. They form a rectilinear network, comprising cracks subparallel to the flow-banded of the lava. (Parallel polarized light)

Fig. 7.6 Rhyolitic breccia outcrop showing angular and cuneiform clasts developing a close-packing arrangement in recrystallized matrix. jigsaw texture is represented by the clasts of intermediate size between the largest ones

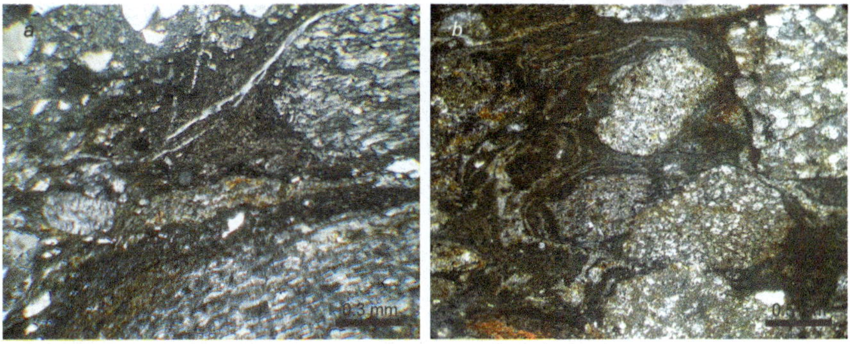

Fig. 7.7 Photomicrographs of rhyolitic breccias, in **a** rhyolitic lava fragments show textural variations like porphyritic or fine microcrystalline quartz–feldspar mosaic. **b** The groundmass develops flow structures contorting around clasts and commonly develops a primary foliation consisting of the alternating glassy and devitrified bands (Crossed polarized light)

Basalts occur at the basal and middle portion of the Carboniferous sequence (e.g., Las Lozas section, Fig. 7.2), as concordant tabular bodies with thickness varying between 2 and 15 m. They are generally massive and porphyritic, also fine-grained. Locally, basalts present flow banding, partly enhanced by the flow orientation of irregular vesicles (2 mm to 1 cm), and filled with chlorites and iron oxides (Fig. 7.8a). Variolitic textures can be observed in some instances. Titan–augite, as main mafic mineral, is commonly fractured and partly altered along fractures, to chlorite and opaque phases. Groundmass is green to brownish and the texture is usually ophitic to subophitic (Fig. 7.8b, c). Glass in the groundmass is altered to palagonite. Vesicles are zoned and filled with chlorite and albite.

The basaltic lava flows are also displaying autoclastic facies at their margins. Clasts, sometimes with altered margins, are angular, ranging among 3–5 cm, rarely reaching up to 60 cm. These autoclastic facies are developed as a response to locally high strain rates in an already more brittle lava body, creating fragments, especially at the outer border of the flows, continually being dislodged by the lava movement (e.g., McPhie et al. 1993).

Fig. 7.8 Photomicrographs showing **a** porphyritic basalt with plagioclase phenocrysts and lesser amounts of augite immersed in a groundmass of plagioclase microlites, augite grains, apatite, and opaque. **b** Basalt with ophitic to subophitic texture. **c** Variolitic texture. (Crossed polarized light)

Fig. 7.9 Volcaniclastic breccias, in **a** clast-supported and poorly sorted breccia composed of angular blocks of rhyolites and rhyolitic vitrophyres. **b** Matrix-supported breccia, angular to subangular clasts of rhyolites, and lesser amount of basaltic ones. **c** Breccia showing angular clasts of recrystallized lavas

7.2.2 Volcanoclastic Lithofacies

They are polymictic volcanic breccias conforming extensive tabular bodies that represent the thickest members in the stratigraphic column (e.g., Las Lozas succession, Fig. 7.2) and are almost entirely composed of lava clasts. Thickness is up to 60 m and they have relatively sharp basal and upper contacts. They are massive but locally show a flow-oriented structure, evident through aligned clasts and internal flow banding (Fig. 7.9a, b). Breccias are very poorly sorted, with clast proportions between 20 and 50% of the rock. Clasts are 1–5 cm in size and may reach up to 60 cm. The beds show increase in the clast size in the upper parts of the tabular bodies. Lithic angular to subangular components are usually predominant, composed mainly of rhyolitic lava fragments with minor proportions of basalt clasts (Fig. 7.9c). Clast proportion in these beds usually increases toward the top.

Thin sections show banded structure (glass or oxidized fragments flowing around lithic fragments), sometimes presenting a foliation constituted by recrystallized quartz and feldspar aggregates, with granophyric textures or with spherulites. Crystal fragments (5–10%) are formed by subhedral fractured feldspars and quartz, up to 3 mm in size. Lithophysae (<1%) are also observed, composed by concentric shells of finely crystalline quartz and feldspars. Irregular vesicles, filled with quartz and iron minerals, are also present in the breccias (Fig. 7.10a, b, c).

Finally, the association of acid members, porphyritic and vitrophyric lavic rhyolites with their equivalent breccia lithofacies (autoclastic lavas; monomictic rhyolite and vitrophyre breccias) is consistent with an eruption mechanism of central lava domes. Consisting of a massive and coherent central part and crumbling outer domains, all mobilize and give form to extensive flows that generate the deposition of large tabular bodies of intermingled coherent and brecciated members. The dominance of rhyolitic lithic clasts in the polymictic breccias lithofacies suggests explosive eruptions that destroyed rhyolite lava domes. In some cases, the breccia facies may be inferred that are above or very close to their source vents. The matrix of these breccias showing spherulitic and granophyric textures, flow-banded compositional structures with abundant glass, and perlitic fragmentation may suggest high temperatures in the environment during their development (c.f. Allen et al. 2008).

The association of rhyolites with their volcanoclastic breccias followed by minor basaltic and intermediate lavas suggests episodic eruption pulses triggering instabilities in the volcanic edifices and degradation of their flanks. The emission of lava and the remobilization of the material, breccia flows, were concomitant with the collapse of the central lava domes. In the studied area, vent sources for these lavas were not yet identified. By other terranes, Allen and McPhie (2002) suggested that widespread outflow would be promoted by eruption from several vents along a fissure rather than from point-source vents, as well it was demonstrated in flood basaltic provinces (e.g., Self et al. 1997).

7.3 Lava Deformation Structures

Identification of deformational structures in lavas is a debatable point, due to which volcanic rocks usually show flow primary textures sometimes very similar to other related to stress and easy to confuse. Nevertheless, there are different authors (e.g., Williams and Burr 1994; van Staal et al. 2001) that could define shear structures on volcanic sequences and indicate the tools to recognize them.

Carboniferous volcanic rocks outcrops recognized along the Pastos Largos Valley (Fig. 7.1) were locally affected by a shear deformation episode (Cisterna et al. 2006), reflected on their structures. Clear signs of this deformation are defined on several volcanic flows that outcrop along a narrow zone and attest the activity of a ductile shear episode. Such deformation is usually maximized by the presence of glass groundmass and/or by the hydration conditions of the lavas (Stünitz and Fitzgerald

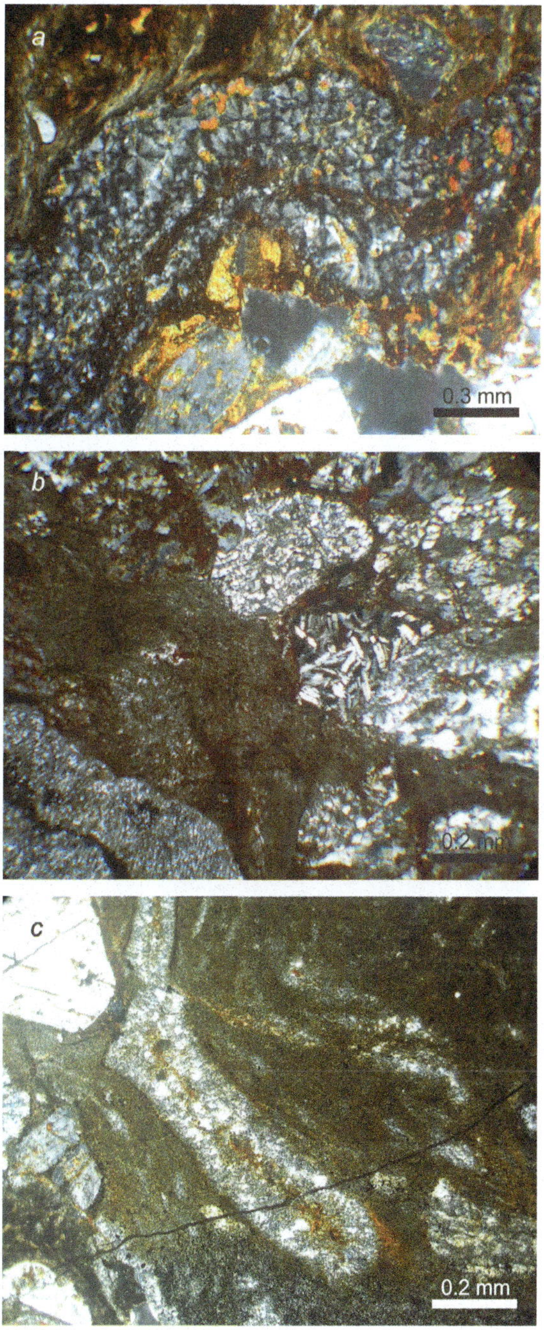

Fig. 7.10 Photomicrographs showing **a** banded structure in polymictic breccias with flow orientation around lithic fragments. **b** Polimictic breccia with clasts of rhyolitic lavas and minor basalt clasts. **c** Lithophysae (<1%) are also observed in the rhyolitic breccia. It is composed of concentric shells with finely crystalline quartz and feldspars. The irregular vesicle is filled with quartz and iron minerals. (Crossed polarized light)

1993; van Staal 1994; van Staal et al. 2001), referred to hydrothermally altered groundmass with finely crystalline quartz, white mica, chlorite, opaque minerals, and minor epidote.

Deformation recognized on rhyolitic breccias is defined by the flattening of the clasts and the spherulites (Fig. 7.11a, b) and by the development of a foliation better to observed in lavas, with the presence of typical shear structures such as sigmoidal shapes and the stretching of phenocrysts along the main foliation (Fig. 7.11c).

The diagnostic shear stress deformation structures are mainly observed at the microscopy scale, where quartz, feldspar, and opaque minerals porphyroclasts

Fig. 7.11 **a** Rhyolitic breccia showing flattened clasts, some of them developing boudins, resulting from a plastic deformation process. **b** Vitrophire with deformed spherulites. **c** Andesitic lava with feldspars defining sigmoidal structures parallel to the main foliation

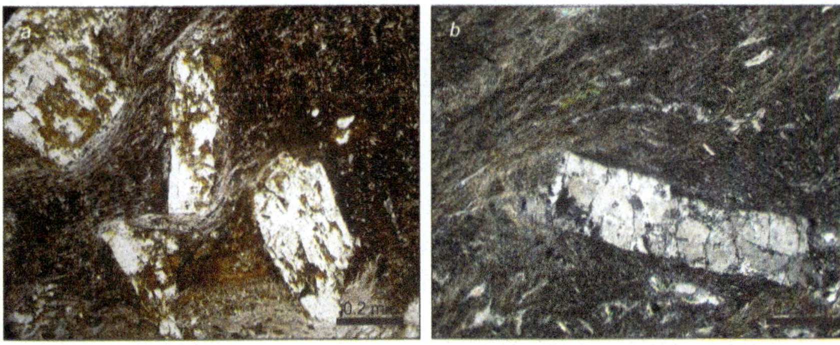

Fig. 7.12 Photomicrographs of **a** breccia showing rotated porphyroclasts with pressure shadows. **b** Plagioclase rotated porphyroclast in a foliated groundmass where are alternating glassy and devitrified bands. (Crossed polarized light)

commonly rotated and surrounded by shadow areas or strain fringes can be recognized (Figs. 7.12a, b and 7.13a, b). The fine groundmass shows foliation structure, usually anastomosed, that locally develops intrafoliated folds (Fig. 7.13c). Biotite fish as lens-shaped single crystals are also common (Fig. 7.13d) as well as boudinage structures on quartz–feldspar veins and quartz and feldspar porphyroclasts (Fig. 7.13e). Chessboard texture in plagioclase and bookshelf structures, both synthetic and antithetic (Fig. 7.13f, g) and deformation lamellae are common in quartz, with undulose extinction or are recrystallized into polycrystalline ribbons (Fig. 7.13h).

The diagnostic structures defined by the lavas and volcaniclastic facies of the Carboniferous sequence provide the data to point a shear deformation episode that generated a mylonitic belt (Cisterna et al. 2006). Its development was related with the input of a fluid phase, enabled to contribute in the deformational process with temperatures relatively low (less than 450 °C) (Cisterna et al. 2006). Such aqueous phase allowed, for example, the dynamic recrystallization of the quartz and albitic feldspar (c.f. Stünitz and Fitzgerald 1993). It may be the resulting of the mobilized fluids from deepest zones in volcanic terranes where there are intercalated sedimentary levels subject to a devolatilization process and transport along deep shear zones (c.f. Bebout and Barton 1989; Selverstone et al. 1991).

0.2 mm
0.3 mm
0.2 mm
0.1 mm

◂**Fig. 7.13** Photomicrographs of **a** Rotated porphyroclasts of quartz and plagioclase forming asymmetric pressure shadows and surrounded by recrystallised quartz. **b** Quartz porphyroclast showing symmetric pressure shadows in the groundmass of a rhyolitic breccia. **c** Intrafoliated folds with axial plane parallel to the main foliation. **d** Biotite fish elongated as a lens at a small angle to the mylonitic foliation. **e** Boudin structure displaying in feldspar, with a typical oblate, lozenge shape, and antithetic microfaults in quartz. **f** Fragmented quartz porphyroclast with undulose extinction. **g** Chess-board texture in feldspar showing a ð-type geometry in a fine groundmass showing quartz–feldspar recrystallization around the porphyroclasts. **h** Boudin of quartz phenoclast showing parallel lamellae development due to intracrystalline deformation (Crossed polarized light)

References

Allen SR, McPhie J (2002) The Eucarro Rhyolite, Gawler range volcanics, South Australia: A >675 km^3 compositionally zoned lava of Mesoproterozoic age. Geol Soc Am Bull 114(12):1592–1609

Allen SR, McPhie J, Ferris G, Simpson C (2008) Evolution and architecture of a large felsic Igneous Province in western Laurentia: The 1.6 Ga Gawler Range Volcanics, South Australia. J Volcanol Geotherm Res 172(1–2):132–147

Bebout GE, Barton MD (1989) Fluid flow and metasomatism in a subduction zone hydrothermal system: Catalina schist terrane, California. Geology 17:976–980

Cisterna CE, Coira B, Koukharsky M (2006) Deformación dúctil en el Valle de Pastos Largos, extremo noroccidental del Sistema de Famatina. Revista De La Asociación Geológica Argentina 6:149–154

Cisterna CE, Coira B, Koukharsky M, Zappettini E, Santos JO (2013) Volcanismo carbonífero pos-orogénico en el río de Las Lozas, valle de Chaschuil, Catamarca, Argentina: nuevas evidencias de extensión en el borde occidental de Gondwana. In: Simposio Argentino del Paleozoico Superior, No. 6, Actas 1: 8. Buenos Aires

Coira B, Cisterna CE, Ulbrich H, Cordani H (2016) Extensional Carboniferous magmatism at the western margin of Gondwana: Las Lozas valley, Catamarca Argentina. Andean Geol 43(1):105–126

Cuerda AJ (1965) Estratigrafía de los depósitos neopaleozoicos de la sierra de Maz (provincia de La Rioja). In: Jornadas Geológicas Argentinas, No. 2, Actas 3, pp 79–94. Salta

Lofgren G (1971) Experimentally produced devitrification textures in natural rhyolite glass. Geol Soc Am Bull 82:553–560

Martina F, Viramonte JM, Astini RA, Pimentel MM, Dantas E (2011) Mississippian volcanism in the south-central Andes: New U-Pb SHRIMP zircón geochronology and whole-rock geochemistry. Gondwana Res 19(2):524–534

McPhie J, Doyle M, Allen R (1993) Volcanic Textures: a guide to the interpretation of textures in volcanic rocks. Centre for Ore Deposits and Exploration Studies, University of Tasmania, Hobart, Tasmania, 198p

Murphy MT, Marsh BD (1993) Textures and magmatic evolution of intermediate-composition dome complexes: evidence from the northern Tatoosh complex, southern Washington Cascades. J Volcanol Geoth Res 54:197–220

Self S, Thordarson T, Keszthelyi L (1997) Emplacement of continental flood basalts flows. In: Mahoney JJ, Coffin M (eds). Large igneous provinces. American Geophysical Union 100, pp 381–410

Selverstone J, Morteani G, Satude JM (1991) Fluid channeling during ductile shearing: transformation of granodiorite into aluminous schist in the Tauern Window, Eastern Alps. J Metamorph Geol 9:419–431

Stünitz H, Fitz Gerald JD (1993) Deformation of granitoids at low metamorphic grade. 11 Granular flow in albite- rich mylonites. Tectonophysics 221:299–324

Turner JM (1960) Estratigrafía del tramo medio de la Sierra de Famatina y adyacencias (La Rioja). Academia Nacional De Ciencias 42:77–126
van Staal CR (1994) The Brunswick subduction complex in the Canadian Appalachians: record of the Late Ordovician to Late Silurian collision between Laurentia and the Gander margin of Avalon. Tectonics 13:946–962
van Staal CR, Rogers N, Taylor BE (2001) Formation of low- temperature mylonites and phyllonites by alkali- metasomatic weakening of felsic volcanic rocks during progressive, subduction- related deformation. J Struct Geol 23:903–921
Williams ML, Burr (1994) Preservation and evolution of quartz phenocrysts in deformed rhyolites. J Struct Geol 16(2):203–221
Zappettini E, Coira B, Santos J, Cisterna CE, Belousova E (2015) Combined U-Pb and Lu-Hf isotopes from the Las Lozas volcanics, Chaschuil Valley, NW Argentina: evidence of lower Pennsylvanian extensional volcanism in western Gondwana. J S Am Earth Sci 59:13–18

The manufacturer's authorised representative in the EU is Springer Nature Customer Service Centre GmbH, Europaplatz 3, 69115 Heidelberg, Germany. If you have any concerns regarding our products, please contact ProductSafety@springernature.com

Printed and bound by CPI Group (UK) Ltd, Croydon, CR0 4YY
30/09/2026
02225469-0006